TRUTH OR BEAUTY

Science and the Quest for Order

DAVID ORRELL

Yale UNIVERSITY PRESS
New Haven and London

First published in the United States in 2012
by Yale University Press.

First published in Canada in 2012
by Oxford University Press.

Yale University Press books may be purchased in
quantity for educational, business, or promotional use.
For information, please e-mail sales.press@yale.edu
(U.S. office) or sales@yaleup.co.uk (U.K. office).

Printed in the United States of America.

Library of Congress Control Number: 2012941261

ISBN 978-0-300-18661-1 (cloth: alk. paper)

A catalogue record for this book is available
from the British Library.

This paper meets the requirements of
ANSI/NISO Z39.48-1992
(Permanence of Paper).

10 9 8 7 6 5 4 3 2 1

For Beatriz,
my architect

Contents

Acknowledgments

Thanks to:

Jennie Rubio and Joseph Camalia for their enthusiastic support of the project and incisive editorial advice.

Peter Woit and Robert Smith for providing valued comments on the manuscript.

My agent Robert Lecker for his guidance over the years.

My former colleagues at the Superconducting Supercollider Laboratory, particularly Jay Jayakumar, Jon Turner, Chris Haddock, Greg Snitchler. We didn't find any new particles, but we shared some good stories.

My parents, who showed me in different ways that numbers and words do mix.

My wife and family for being both beautiful and true.

Introduction

"Beauty is truth, truth beauty,"—that is all
Ye know on earth, and all ye need to know.

John Keats, "Ode on a Grecian Urn"

My work always tried to unite the true with the
beautiful, but when I had to choose one or
the other, I usually chose the beautiful.

Hermann Weyl, quoted in Freeman Dyson,
"Obituary of Hermann Weyl"

Behind it all is surely an idea so simple, so
beautiful, that when we grasp it—in a decade,
a century, or a millennium—we will all say to
each other, how could it have been otherwise?

John Archibald Wheeler, *How Come the Quantum?*

The machine known as the Large Hadron Collider (LHC) is located in a quiet, pastoral site adjacent to the Jura Mountains, near Geneva. Cows munch contentedly in green fields; a number of very graffiti-free low-rise buildings dot the horizon. On the surface, there is little to suggest that the area hosts one of the most ambitious art/science fusion projects in history.

The installation is located down below, about 100 meters underground, in a round, twenty-seven-kilometer tunnel that straddles the border of Switzerland and France. The tunnel is packed with an almost continuous chain of over a thousand superconducting magnets, which steer twin beams of protons traveling in opposite directions at near the speed of light. The beams are made to collide in the tunnel's four experimental halls, which are known by their acronyms: ALICE, ATLAS, CMS, and LHCb. The names are unexciting, but apart from their invitation-only inaccessibility to the public, these caverns are the cathedrals of the modern age—a heady mix of advanced engineering and something approaching religion. They are the location where the LHC's users hope to find beauty and a connection with eternity.

The ATLAS detector (A Toroidal LHC ApparatuS), for example, is a collaboration of over 1,800 people from thirty-four countries. Its massive octagonal structure, as high as a five-storey building and weighing 7,000 tonnes, is packed with complicated lattices of steel, ceramic, and plastic, all connected by conduits and thick bundles of cables. It's an awe-inspiring spectacle, like something from a science-fiction fantasy. Despite its huge size, the purpose of its 100 million sensors is to analyze the smallest things man has ever produced—the stream of subatomic particles emitted when the opposing beams silently collide in a tiny cauldron of intense energy. Like the world's largest and most expensive digital camera, the detector snaps off millions of images every second in the hope of capturing exotic events and phenomena such as the Higgs field and its associated boson (the recently discovered particle thought to be responsible for giving matter its mass), hypothesized "supersymmetric" particles, and perhaps even very small black holes.

Data from such happenings are recorded and streamed out to universities and research institutions around the world, where they are analyzed by scientists. Poring over the numbers and pondering the complex images, they hope to gain insight into profound questions. Are there extra dimensions of space? Why is there more matter than antimatter? What is "dark matter" composed of? What happened in the first moments of creation? And what will happen in its last?

But what really drives and excites these researchers is a quest for something even more elusive, desirable, and beautiful.

For centuries, scientists and philosophers have striven to show that the universe is governed by a few simple principles. These are not physical rules or components of matter.

They are deeper even than mathematics.

They are aesthetic laws, based on concepts such as harmony, unity, and symmetry.

Scientists aren't just expecting a picture of a particle. They want an image of the truth.

Where Art and Science Collide

One does not normally think of science as being an aesthetic pursuit, nor of particle accelerator projects like the LHC as highly expensive art installations. The whole point of science, after all, is to be objective and impartial. Whether a theory or experiment is in some sense ugly or attractive should not enter the calculation. All that matters is whether it works. Beauty is the province of (non-modern) art museums.

It is certainly the case that, if you peruse scientific journals, you will see little mention of appearances or aesthetics or elegance. No one claims that their theory is superior because it is better looking than its competitors. Scientific discoveries made at the LHC are not rated by a beauty contest. But if you listen to scientists talking, or read what they write outside of peer-reviewed articles, then a very different picture emerges: there is a general acceptance that beauty and truth are mysteriously and inextricably linked. Indeed, the central drive of science often seems to be as much a quest for beauty as for truth, on the understanding that the two are to be found in the same place.

Bertrand Russell wrote that "mathematics, rightly viewed, possesses not only truth, but supreme beauty—a beauty cold and austere, like that of sculpture, without appeal to any part of our weaker nature, without the gorgeous trappings of painting or music, yet sublimely pure, and capable of a stern perfection such as only the greatest art can show."[1] In his 1940 book *A Mathematician's Apology*, G.H. Hardy wrote that "The mathematician's patterns, like the painter's or the poet's must be *beautiful*; the ideas like the colours or the words, must fit together in a harmonious way. Beauty is the first test: there is no permanent place in the world for ugly mathematics."[2] The French mathematician and physicist Henri Poincaré believed that this "feeling of mathematical beauty, of the harmony of numbers and forms and of geometric elegance . . . [this] real aesthetic feeling that all true mathematicians recognize" acted as a "delicate sieve" which allowed the scientist to discern the truly useful patterns in nature.[3]

The correspondence between mathematics and the physical world, according to the British physicist Paul Dirac, meant that "it is more important to have beauty in one's equations than to have them fit experiment."[4] He proved this by using a very elegant equation to postulate the existence of antimatter before it had been physically detected. When asked to back up his quip that in his work he "usually chose the beautiful" over the truth, the mathematician and theoretical physicist Hermann Weyl mentioned his gauge idea for gravitation, which while wrong for gravity later reshaped other areas of physics as discussed in Chapter 7.[5] In a 1960 essay, physicist Eugene Wigner described the "unreasonable effectiveness" of beautiful mathematics in the natural sciences.[6] The physicist Murray Gell-Mann noted in a recent talk that "what is especially striking and remarkable is that in fundamental physics a beautiful or elegant theory is more likely to be correct than a theory that is inelegant."[7] According to physicist Anthony Zee, the "rallying cry of fundamental physicists" is, "Let us worry about beauty first, and truth will take care of itself!"[8]

This interest in aesthetics extends even to the designers of facilities. When Robert Wilson, who was the director of the Fermi National Accelerator Laboratory (Fermilab) in Illinois, was called in front of Congress in 1969 to defend the project, he stated that its value had nothing to do with national security: "It only has to do with the respect with which we regard one another, the dignity of men, our love of culture . . . it has to do with: Are we good painters, good sculptors, great poets? I mean all the things that we really venerate and honor in our country and are patriotic about . . . it has nothing to do directly with defending our country except to help make it worth defending."[9] He later backed up his words by contributing to the architectural design of Fermilab, including the splendid main building which was modeled after a cathedral in Beauvais, France.

Wilson's statement to Congress was perhaps a little disingenuous; despite his protestations, in the height of the Cold War it was widely assumed that high-tech projects like accelerators had everything to do with national security, which is why they attracted so much funding. Even when I worked on the design of a (later canceled) successor to Fermilab called the Superconducting Super Collider in the 1990s, the project had a quasi-military feel to it, with much of the work carried out by defense contractors. However, Wilson's words still have a high degree of truth, and certainly anyone hoping to get a death ray out of the Large Hadron Collider will likely be disappointed.

Artists and musicians, in return, have long exploited mathematics and proportions in their work. In his 1926 book *Transformations*, the

English art critic Roger Fry compared the emotional states produced by contemplating art works and mathematics, concluding that "it would be impossible to deny the close similarity of the orientation of faculties and attention in the two cases."[10] Writing on the aesthetics of music, philosopher Roger Scruton wrote, "When understanding mathematics we have access to the order of creation, and this order is eternal, like the numbers themselves. In music we know through experience, and in time, what is also revealed to the intellect as outside time and change."[11] Art and science are two ways of exploring reality that are driven by similar impulses and can evoke the same kind of wonder, excitement, or comfort.

The Universe on a T-shirt

Just as we ascribe laws to nature, so we have our laws of aesthetics (though as discussed later, these may vary with time and place). Three classic aesthetic properties, which were celebrated by the ancients and apply to mathematical formulae and scientific theories as well as they apply to art or architecture, are elegance, unity, and symmetry. As Stephen Hawking told the *Guardian* newspaper, "Science is beautiful when it makes simple explanations of phenomena or connections between different observations. Examples include the double helix in biology, and the fundamental equations of physics."[12] Physicist Steven Weinberg notes that "when we formulate the equations of quantum field theories or string theories we demand a great deal of mathematical elegance, because we believe that the mathematical elegance that must exist at the root of things in nature has to be mirrored at the level where we are working."[13] Perhaps the

Figure 0.1 Image by artist Laurent Mulot, from a series of images he calls *Augenblick* (German for "instant"). The top panel is a photograph taken in the area above the Large Hadron Collider. The bottom image is the computer-generated output from a collider experiment which occurred at the same time as the photograph. Source: Laurent Mulot

archetype of a beautiful theory is Newton's law of gravity, which computes the gravitational force between any two objects.

Newton's equation is elegant because it involves only the masses of the objects, the square of the distance between them, and a single parameter called the gravitational constant. It doesn't matter what color the objects are, what they are made of, or what they look like. All those details are mere clothing and can be stripped away by the principle known as Ockham's razor, which was well expressed in a phrase attributed to Einstein: "Everything should be made as simple as possible, but not simpler."[14] In science, the idea of beauty is therefore tied up with reductionism.

The law of gravity is powerfully unifying, because it explains a wide range of disparate phenomena—from the arc of a cannonball to the motion of the Moon around the Earth—using a single formula. And it is circularly symmetric, in the sense that the gravitational pull exerted by an object will be the same at any point an equal distance away. The Earth is round because gravity pulls its mass uniformly towards its center. As we will see, the search for symmetry has played a defining role in science, from the Greek models of the cosmos to modern theories of supersymmetric particles.[15]

Newton's success set a high standard for other physicists and inspired them to find similar formulae for other phenomena. The most famous examples are Maxwell's equations, which relate electricity, magnetism, and the speed of light; and Einstein's $E=mc^2$, which equates energy and mass. The ultimate aim of physics is the Theory of Everything: a set of equations which describe the entire universe, including all forces and forms of matter, in just a few lines of mathematics. Again, the central drive towards this theory appears to be based on a kind of sparse aesthetics. Citing the "symmetry, simplicity and beauty that can be described by abstract mathematics," physicist Leon Lederman argues that the theory should ideally comprise "a formula so elegant and simple that it will fit easily on the front of a T-shirt."[16]

A physicist's notion of beauty is clearly a little different from that usually encountered in the arts, especially in the way that it aims for a final, idealized, "true" solution. There is ultimately no room for alternative schools or different cultural influences. As physicist John D. Barrow puts it, "The longed-for Theory of Everything promises to provide the final discovery after which all physics will become the refinement of its content, the simplification of its explanation. . . . Eventually, it will appear on T-shirts."[17] The MIT cosmologist Max Tegmark told *New Scientist* that "in 2056, I think you'll be able to buy a T-shirt on which are printed equations describing the unified physical laws of our universe."[18] (This interest in T-shirts may actually say more about physicists' sartorial habits than the underlying structure of the cosmos.)

Because other sciences tend to model themselves after physics, the discipline's emphasis on elegance, simplicity, and reductionism have had a huge influence on them as well. In biology, the role of atoms is played by genes; in the social sciences, it is played by atomistic individuals. Perhaps the strongest example is economics, where assumptions such as the stability of markets and the rationality of market participants have allowed the creation of a mathematically elegant pseudo-physics of human behavior.

The biologist and renowned science communicator Richard Dawkins describes the "feeling of awed wonder that science can give us" as "a deep aesthetic passion to rank with the finest that music and poetry can deliver."[19] Given the evident power of aesthetics as a guide to scientific sensibility, it might therefore seem a little unappreciative to call its value into question—like suggesting that the Taj Mahal is pretty but not the most efficient use of space. But in this book I will argue that, while the scientific quest for beauty has motivated and driven generations of scientists, it may in recent years have led us down the wrong track—and not for the first time.

Science: A Love Story

The idea that a preoccupation with aesthetic principles such as symmetry or unification might, if misapplied, be harmful to science is not entirely new. In *The Sleepwalkers* (1959), Arthur Koestler criticized the Platonic tendency to "think in circles," which he saw as prevalent in science.[20] Fritjof Capra's *The Tao of Physics* (1975), which hit a powerful chord with a large public if not all physicists, argued that the findings of particle physics were more in tune with ideas of beauty from Eastern philosophy than with classical Western ideals.[21] In his 1993 book *The End of Physics*, David Lindley wrote that the search for a unified physical theory of "compelling beauty" in which "truth will be beauty and beauty will be truth" was unlikely ever to be achieved.[22] This was followed in 1996 by John Horgan's *The End of Science*, which argued that a lack of empirical data could mean that much scientific work would end up being judged primarily on the basis of aesthetics.[23] Roger Penrose's *The Road to Reality* (2004), which aimed to present the equations that describe the universe to a non-specialist audience, highlighted throughout "the remarkable role that aesthetic judgements play" in scientific research, but warned that "mathematical beauty is, by itself, an ambiguous guide at best."[24]

Robert Laughlin, in *A Different Universe* (2005), questioned the relevance of "models that were beautiful but predicted no experiments" and described symmetry as "an important, if often abused, idea in physics."[25] In *The Trouble With Physics* (2006), theorist Lee Smolin pointed out the drawbacks of modern physics, particularly the string theory program, and agreed that "mathematical beauty can be misleading."[26] The same year Peter Woit's *Not Even Wrong* argued that actually one of the problems with string theory is that it is not beautiful enough, in the sense that it does little to simplify or unify the laws of physics.[27]

In my 2007 book *Apollo's Arrow*, I wrote that our mathematical models used for predicting the future had been shaped by what amounted to a deeply-held "set of aesthetic principles" which were not always compatible with reality, and argued for a more balanced approach.[28] In *The Boundaries of the New Frontier* (2009), rhetorician Joanna S. Ploeger showed how an appeal to aesthetics can be used as a kind of public relations strategy by scientists to promote and protect their interests.[29] Marcelo Gleiser's *A Tear at the Edge of Creation* (2010) asserted that the structure of the cosmos is the result of imperfections, imbalances, and asymmetries, and that we need to develop an "aesthetics of the asymmetric" that acknowledges imperfection.[30] In 2012, when the scientific website Edge.org surveyed its contributors with the question "what is your favorite deep, elegant, or beautiful explanation?" the result was a diverse range of responses, some of which are quoted here.[31] *Truth or Beauty* concentrates less on expounding a vision of the universe, or promoting a particular theory, than on the scientific sense of beauty as a subject in its own right. We examine the topic from a number of different angles, explore its history and sociology, show how it has affected fields from cosmology to meteorology, and illustrate in concrete terms how methods from areas such as complexity theory are pointing the way to a new aesthetic.

The book is divided into three sections. These chart the relationship between science and beauty over three stages: infatuation, complications leading to crisis, and a shift towards a kind of maturing—in which, like an aging Lothario, we may be on the verge of admitting that looks don't count for everything. Individual chapters feature particular aesthetic themes, such as unity or symmetry, interleaved with excerpts from the history of science that illustrate the scientific sense of beauty and highlight its role in the development and interpretation of scientific theories.

The first section (Chapters 1 to 3) gives an overview of the role of beauty and aesthetics in science from the time of the ancient Greeks to the early twentieth century. The ancients constructed a cosmology based on the idea of musical harmony, which was related to numerical ratios. The planets were believed to produce musical tones as they moved around the Earth. Together with the Moon and stars, they produced a beautiful music—what Pythagoras called the Harmony of the Spheres. If the cosmos was based on numbers, it followed that art and architecture should be as well. Temples were thus constructed based on the same mathematical ratios that described musical harmonies.

The Greek cosmological model persisted for centuries and was not overturned until the Renaissance, when scientists including Kepler, Galileo, and Newton based their theories on a close examination of nature. Newton believed that matter was made up of "solid, massy, hard, impenetrable, movable particles" governed by physical laws—an idea that went back to the atomic theory of Democritus. His "rational mechanics" unified physics over all scales, from the atom to the cosmos, and laid out a program that science would follow for the next few hundred years: reduce a system to its fundamental components, determine the relevant physical laws, express them as mathematical equations, and solve.

The formal test of a physical theory is not whether it is pretty, but whether it can make accurate predictions. By this criterion, as discussed in Chapters 4 to 7, the mechanistic approach proved tremendously successful in many areas, particularly the "hard" sciences such as physics and chemistry—even managing to survive, in a modified form, the considerable challenges thrown up by quantum physics. It was aided in the twentieth century by technological advances including particle accelerators, which allow scientists to probe the inside of atoms; and the terrible beauty of the atom bomb, which, for all its horror, meant that huge resources were poured into nuclear research. By the late 1960s, scientists had devised a mathematical set of laws, known as the Standard Model, that appeared to accurately simulate subatomic experiments, and this model remains in place today. However, the model has serious deficiencies. As Steven Weinberg, who gave the theory its name, puts it, "we know that the Standard Model is not the final answer, because of its obvious imperfections—and those imperfections, I have to say, are aesthetic."[32] (Viewed in this way, the LHC, which is designed to address those limitations, really is an art project.)

The main aesthetic drawback is excessive complication. While the Standard Model does manage to unify two of the Earth's known forces (electromagnetic and weak nuclear), it still requires some twenty parameters to be arbitrarily set. For comparison, Newton's law of gravity gets by with just one such parameter: the gravitational constant. The theory also fails to properly accommodate a number of phenomena, including gravity. In recent decades, physicists have therefore focused on coming up with a replacement for the Standard Model.

Today, the prime candidate for a Theory of Everything is string theory, which claims that fundamental particles such as electrons are not particles at all, but tiny strings oscillating in a ten-dimensional space (nine spatial dimensions, only three of which we can see, plus time). This theory is widely considered by its practitioners to be the ultimate example of a beautiful theory, because of its deep symmetries and potential explanatory power.

However, while string theory is based on a beautiful idea—little vibrating strings—its critics point out that it enjoys no empirical support. Its predictions are untestable except at extremely high energies, which cannot be produced by even the largest particle accelerator. Even worse is the fact that it would be almost impossible to validate by any experiment: the theory in its current state is so general, and exists in so many different variations, that it could be adjusted to fit just about any findings. Its beauty refuses to be pinned down. Many observers therefore believe that physics is in a state of crisis. So, do we need a larger collider—or have we been led astray by the search for a particular type of beauty?

Mental Models

The final chapters of this book ask whether our difficulties in modeling the cosmos or human society are part of a larger pattern, and whether we have become too focused on a narrow class of problems at the exclusion of much else. Have we become so entranced by a certain kind of beauty that we obsessively follow it everywhere, ignoring its failings? That we even hallucinate about its presence when it isn't there? Have we crossed the line between appreciating beauty and stalking it?

Human beings have long been trying to explain the natural world by reducing it to number. The metaphor, since at least the seventeenth century, has been of the world as a beautiful machine.[33] To understand the machine, all we need is to take it to pieces—for example by smashing protons together at high speed—and figure out how they relate. But the approach isn't working as well as it once did. In many respects, it seems that the phenomena we are dealing with are best seen not as part of an elaborate machine but as part of a complex organic whole.

A defining property of complex systems is that they exhibit what is known as emergent behavior: properties which emerge from the system but cannot be predicted using knowledge of the system's components alone. This throws a spanner in the Newtonian works and calls for a fundamentally different approach. As physicist Robert Laughlin observes, "Science has now moved from an Age of Reductionism to an Age of Emergence, a time when the search for ultimate causes of things shifts from the behavior of parts to the behavior of the collective."[34] It also calls for a new aesthetics, for a consequence of this shift is that we will no longer look for neat Theories of Everything which unify phenomena over all scales. Instead, models are more like patches which reveal a portion of the complex whole.

The crisis in physics also extends to other sciences which have been strongly influenced by it. In biology, the theory of the selfish gene provided a kind of atomic analogue for life in which our lives are determined by the beautiful spiral of our DNA. Neoclassical economics built an entire worldview around the idea that beautifully rational individuals and firms behave like independent atoms, optimizing their own utility in the marketplace. Today, the critique of economics resembles that of string theory: the theory might be mathematically elegant, but it is of little use in making practical predictions, as shown by the failure to foresee the crash of 2007/2008.

It might seem that mathematical theories or experiments on subatomic particles have little impact on our lives, but what is certainly true is that the scientific aesthetic, which informs and motivates those theories and experiments, has played a huge role in shaping our mental and material worlds. Everything from architecture to military weaponry, to the way we treat disease, to the structure of the economy has been influenced by the scientific aesthetic.

Our changing approach to physics gives us a new set of metaphors and stories for interpreting physical reality, and this is already having a tangible effect on things which might appear to be light-years away from theoretical physics. Physicists have always been aware of parallels between their abstract equations and what happens out in the real world. Paul Dirac noted in 1933 the "great similarity between the problems provided by the mysterious behaviour of the atom and those provided by the present economic paradoxes confronting the world."[35] David Bohm warned that the reductionist approach to the world "has led to the growing series of extremely urgent crises that is confronting us today."[36] Our mental models affect the way we see the world, and therefore they affect the real world itself.

To better understand the universe and our place in it, we are going to need more than telescopes and atom-smashers. At the heart of the transformation is a shift in aesthetics, from order and symmetry to something more complex, organic, and messy. The structures to be erected will be fluid and curved instead of square and static. Symmetry and perfection will be seen as special cases, rather than the authors of the universe. We will learn to celebrate qualities such as duality, mutability, and asymmetry—not just in physics or in science, but in our entire world view.

This will have consequences. As the writer J. G. Ballard said, "I suspect that many of the great cultural shifts that prepare the way for political change are largely aesthetic."[37] Art and science are two versions of the same matter—and when they collide, the results are bound to be interesting.

I

INFATUATION

1

Harmony

From harmony, from heavenly harmony,
This universal frame began:
When nature underneath a heap
Of jarring atoms lay.
And could not heave her head,
The tuneful voice we heard from high:
Arise, ye more than dead.

John Dryden, "A Song for St. Cecilia's Day"

Many people think that modern science is far removed from God . . . in our knowledge of physical nature we have penetrated so far that we can obtain a vision of the flawless harmony which is in conformity with sublime reason.

Stephen M. Barr, *Modern Physics and Ancient Faith*

After a while, another voice said: One, two, three, four—And the universe came into being. It was wrong to call it a big bang. That would just be noise, and all that noise could create is more noise and a cosmos full of random particles. Matter exploded into being, apparently as chaos, but in fact as a chord. The ultimate power chord.

Terry Pratchett, *Soul Music*

For centuries, sages, mystics, priests, and more recently scientists have sought an explanation or model for the universe. A common component in many of these descriptions is a fascination with harmony. The Greeks thought that the cosmos was based on musical harmony, and this belief has persisted, in modified form, over the ages. Scientists ranging from Johannes Kepler in the seventeenth century to modern string theorists have used harmony to describe the universe. In this chapter, we put our ear to the ground and to the skies to pick up the timeless melodies that have entranced, inspired, and sometimes confused us since the dawn of science.

The Buddha observed that the question of how the universe was formed would only "bring madness and vexation to anyone who conjectured about it."[1] Nonetheless, most civilizations have given it a try and produced some kind of creation story, an overarching theory that attempts to explain why things are the way they are.

In a Chinese version, for example, the universe began with the opposing powers of Yin and Yang, who produced an offspring called Pan Gu that grew for 18,000 years inside a great egg. When it finally hatched, the dark part (Yin) sank down to form the Earth while the light part (Yang) rose up to form the sky. After another 18,000 years Pan Gu died, and its dismembered body parts formed the Sun, the Moon, the stars, and all the rest of the universe.

In classical Greek mythology, everything began with Chaos, who gave birth to five progeny including Gaia, the Earth goddess. She in turn spawned Uranus, the sky, and Pontus, the sea; and then coupled with her own child, Uranus, to create more gods and goddesses.

Our current version of the creation story reconciles astronomical observations with studies of matter carried out at physics laboratories. According to this, the universe was produced from an infinitesimally small dot in an event known as the big bang.[2] It happened about 13.7 billion years ago and we are still feeling the effects. The universe hasn't stopped expanding. In fact, rather mysteriously, the expansion is accelerating. We can hear the birth pangs of the early universe in the so-called cosmic microwave background, whose detection in 1964 provided tangible support for the big bang theory.

Scientists can only hypothesize about what happened in the earliest moments, but the story goes that all the fundamental forces that make up the universe and hold it together—gravity, electromagnetism, and the weak and strong nuclear forces—were united in perfect symmetry so that all things were one, in a kind of pulse of pure energy. As the universe expanded and cooled, the forces separated and distinguished themselves according to patterns that were programmed by mathematical laws. Recognizable particles such as electrons, protons, and neutrons emerged into being, all colliding together chaotically with tremendous energies. At this point, the universe was about a second old.

After a minute or so, protons and neutrons combined to form atomic nuclei, but the temperature was still some 3 trillion degrees centigrade. It would take hundreds of thousands of years more for the universe to cool sufficiently for nuclei to be able to join with electrons and form complete atoms. It took longer still for complex molecules such as water to appear; and it took about 9 billion years for the planet Earth to form itself from a swirling mass of vapor around the Sun.

Today, the universe has cooled to an average temperature of about -270°C. Most places on Earth are of course considerably warmer, thanks to radiation from the Sun. And then there is the Large Hadron Collider, where scientists—with the occasional break for complicated and time-consuming repairs—have been risking "madness and vexation" since 2008 to recreate, in a controlled way, the ultra-hot conditions that existed shortly after the big bang.

The technology is cutting edge, but its aim is ancient: to hear the flawless, heavenly harmonies that were present at the dawn of time.

Number Is All

The Greek philosopher Pythagoras didn't have a particle accelerator, but he too was preoccupied with the questions of where we came from, what we're made of, and where we are going. He also had a creation myth. According to the Pythagoreans—the quasi-religious cult he established in the sixth century BC—in the beginning there was Unity. It then divided into two components, the Limited and the Unlimited.

These two, which were opposite, came together to form numbers, which the Pythagoreans believed formed the structure of the cosmos (a word coined by Pythagoras).

To the Pythagoreans, each number had its own meaning. One, the monad, represented the initial unified state which marked the creation of the universe—in today's terms, the state at the start of the big bang. Two, the dyad, represented the polarization of unity into duality, and was associated with change, mutability, and the feminine. Three signified all things with a beginning, middle, and end; while four represented completion, as in the four seasons that make up a year. The greatest and most perfect of all numbers was ten, the sum of the first four numbers, which symbolized the universe.

Pythagoras is best known today for his theory concerning right triangles, but his most profound achievement was the discovery that musical notes are governed by mathematical ratios. The story goes that Pythagoras was outside a blacksmith's shop when he noticed that the sounds produced by different hammers were in some cases consonant with one another, and in other cases inconsonant. Intrigued, he entered the shop to investigate and discovered that the hammers which harmonized well together had weights which were related by simple ratios, such as 1:2, 2:3, or 3:4.

As with much of what has been passed down about Pythagoras, this story was clearly made up; in fact the tones produced by hammers do not vary in proportion with weight. The story probably reflects an ancient tradition that associated wizardry and magic with blacksmithing.[3] The idea does work perfectly, however, for the strings of a musical instrument. As the Pythagoreans also showed, a string plucked on an instrument such as a lyre will produce a certain note. Fret the string exactly halfway up (ratio 1:2) and the note raises an octave. Fretting 2/3 of the way up produces a musical fifth, and 3/4 up a fourth. Playing the three notes together on different strings gives a major chord.

These musical ratios of 1:2, 2:3, and 3:4 were represented by the four rows of the tetractys, shown in Figure 1.1. The Pythagoreans considered the tetractys to be a sacred symbol and associated it with both harmony and the Delphic oracle (see Box 1.1). A Pythagorean aphorism read, "What is the oracle at Delphi? The tetractys."[4]

Since music was considered to be the most expressive and mysterious of art forms, the discovery that harmony was ruled by mathematical laws was powerful evidence that the universe could similarly be reduced to numbers. "Greek philosophy," noted the scholar John Burnet, "was henceforward to be dominated by the notion of the perfectly tuned string."[5]

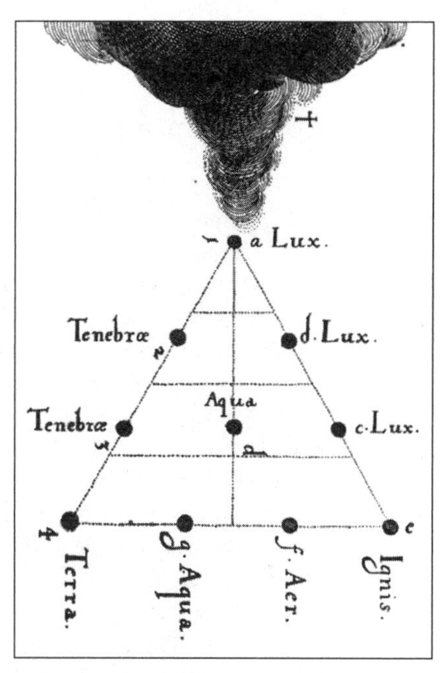

Figure 1.1 A seventeenth-century version of the Pythagorean tetractys by the physician and mathematician Robert Fludd (1574–1637). The ratios between the rows are 1:2, 2:3, and 3:4, which correspond to the musical octave, fifth, and fourth.[6] The drawing by Fludd also illustrates the Pythagorean creation myth. The dark cloud represents the state of chaos before the universe was born. The creation of the universe was marked by the appearance of the monad, or unity, followed by the dyad, followed by the other numbers. Source: Reprinted from Robert Fludd, *Philosophia sacra et vere christiana seu meteorologia cosmica* (Frankfurt: Francofurti prostat in officina Bryana, 1626).

The Harmony of the Spheres

For the Greeks, the word "cosmos" did more than represent the observed universe; as scholar W.K.C. Guthrie notes, it united "the notion of order, arrangement or structural perfection with that of beauty."[7] Central to this was the idea that the universe is based on mathematical harmony. In the same way that musical harmony imposes a kind of limit and order on the unlimited continuum of possible tones, so harmony was the organizing principle that brought together the Limited and the Unlimited when the universe was formed. Its tune was the score for the choreographed movements of the heavenly bodies.

Box 1.1

The Apollo Principle

According to Greek mythology, the oracles at Delphi were originally provided by Gaia, the Earth goddess. The young god Apollo killed Gaia's protector, the serpent called Python, and took over the oracle. From then on, he was known as the Pythian Apollo. He later took over from Helios as the god of the Sun. According to Plutarch, the name "A-pollo" (not many) referred to the concept of unity.

The Greeks associated Apollo with numerous things including light, music, and prophecy. He was also a warrior. His hostile takeover of the oracle was an example of what Joseph Campbell referred to as solarization, "whereby the entire symbolic system of the earlier age is to be reversed, with the moon and the lunar bull assigned to the mythic sphere of the female, and the lion, the solar principle, to the male."[8]

The oracle at Delphi went on to become the most successful forecasting operation in history, lasting for almost a thousand years from the eighth century BC. The predictions were made by a woman known as the Pythia, who was chosen from the local population as a channel for Apollo. One of her more famous predictions was the birth of Pythagoras, who was named after the Pythia.[9] His followers believed that he was a demigod descended directly from Apollo, with superhuman powers such as the ability to dart into the future.[10]

The philosopher Friedrich Nietzsche identified the Apollonian principle in Greek art as corresponding to detached, logical, sculptured representations of the world, which he saw as a way of escaping from messy reality in favor of "beautiful illusions." He contrasted it with the Dionysian principle—Dionysus being the god of wine, among other things—which was wild, instinctual, and rooted in nature, and was best expressed through forms such as music and dance. In *The Birth of Tragedy*, Nietzsche argued that while both were necessary in art, Western culture and science in particular had long been dominated by the Apollonian principle.[11]

James Hillman, who developed post-Jungian archetypal psychology in the 1970s, equated the Apollo archetype with the properties of "detachment, dispassion, exclusive masculinity, clarity, formal beauty, farsighted aim and elitism"—which, as seen below, describe quite well the core values of reductionist science.[12] The highest reward for the scientific elite is to become a Nobel laureate, which is fitting since the symbol of Apollo in ancient Greece was the laurel tree: he was often depicted in artworks with laurels in his hair, and laurels were used as a wreath to honor heroes.

> The dominance of the Apollo principle appeared complete in the late 1960s when, for reasons largely related to Cold War propaganda, spacecraft with the god's name on them traveled to the Moon, thus conquering in a very real sense Campbell's "mythic sphere" of the lunar and the female. Ironically, though, the most interesting things they brought back were not pieces of Moon rock, but pictures of our planet. For the first time, we saw the Earth as a vibrantly living thing against the backdrop of space.[13] Apollo, meet Gaia.

Because ten was the perfect number, it followed to the Pythagoreans that there should be ten heavenly bodies. According to the Pythagorean philosopher Philolaus, these consisted of the Earth, the Moon, the Sun, the five planets that were known at the time, the stars (which were considered a single outer layer of fire), and something called the counter-earth. The Pythagoreans believed that all of these rotated around a "central fire" which, like the counter-earth, was always opposite to us and out of sight.

The Pythagoreans were well ahead of their time in believing that the Earth could be in motion. Their invention of the counter-earth, on the other hand, was an early example of scientists inventing a phenomenon—in this case a planet—in order to meet the requirements of an elegant theory. As Aristotle drily observed, "And all the properties of numbers and scales which they could show to agree with the attributes and parts and the whole arrangement of the heavens, they collected and fitted into their scheme; and if there was a gap anywhere, they readily made additions so as to make their whole theory coherent. E.g. as the number 10 is thought to be perfect and to comprise the whole nature of numbers, they say that the bodies which move through the heavens are ten, but as the visible bodies are only nine, to meet this they invent a tenth—the 'counter-earth.'"[14]

The movement of each of these ten bodies was thought to produce a musical tone, with the pitch depending on the speed of rotation: the fast-moving Sun had a higher pitch than the laggard Moon, for example. Together the tones produced a beautiful music known as the Harmony of the Spheres. As his biographer Iamblichus wrote, only Pythagoras was capable of "hearing and understanding the universal harmony and consonance of the spheres and the stars that are moved through them, which produce a fuller and more intense melody than anything effected by mortal sounds."[15] To the rest of us this celestial concert went unnoticed; having been exposed to it since birth we had become unaware of it, like people who live next to a railway track and have learned to tune out the sound of trains.

The Pythagorean assertion that the universe was ordered and rational and based on mathematical harmony had far-reaching and radical consequences, for it implied that there was a kind of sympathy between our faculty of reason and the cosmos. As the philosopher Sextus Empiricus wrote, "The Pythagoreans say that reason is the criterion of truth—not reason in general, but mathematical reason."[16] Einstein later identified this belief in rationality as akin to religion: "I have no better expression than 'religious' for confidence in the rational nature of reality insofar as it is accessible to human reason."[17] By exercising reason, we should therefore be able to understand the universe—and, as physicist Stephen Hawking put it, "If you understand how the universe operates, you control it in a way."[18] Instead of being at the whim of random events, or of squabbling and recalcitrant gods, we could be masters of our own destiny.

The Standard Model

If the cosmos was based on number, it followed that the movements of the heavenly bodies could be modeled and predicted using mathematics. In a time when events in the human sphere were believed to be strongly influenced by astrology, this was of no small practical interest. The Greeks therefore set about building sophisticated geometric constructions which could be used to chart the skies.

These early versions of mathematical models were based on two main assumptions: that everything moved around the Earth (the notions of the central fire and counter-earth did not survive); and that everything moved in circles, since circles and spheres were the most symmetric—and therefore the most pure—of forms. In his *Republic*, for example, Plato described (from inner to outer) the Moon, Sun, Venus, Mercury, Mars, Jupiter, Saturn, and stars as rotating in concentric circles: "on the upper surface of each circle is a siren, who goes round with them, hymning a single tone or note. The eight together form one harmony."[19]

The assumption of circular motion around the Earth worked well enough for the Moon, Sun, and stars. The planets, however, followed a more complicated path. Mars, for example, would complete a full revolution in about 780 days, but partway round it would perform a retrogradation lasting about ten weeks in which it slowed down and backtracked a bit before proceeding—like an unruly soloist in a jazz orchestra.

In the third century BC, Plato's associate Eudoxus came up with a way to reconcile this motion with Pythagorean spheres. Each body was imagined to be located on a sphere, rotating around an axis whose center was the Earth. To account for effects such as retrograde motion, the endpoints of the axes of rotation were fixed on other spheres that themselves rotated. This allowed for highly complex motion. The complete model of twenty-seven spheres was a remarkable achievement in mathematical modeling, and it captured the overall motions of the cosmos even if it could not give a perfect fit to the observed data.

Other astronomers, including Plato's former student Aristotle, later improved on the model by adding more spheres, which allowed for better reproduction of planetary motion. It was again believed that the spheres produced musical tones as they rotated. "It seems that bodies so great must inevitably produce a sound by their movement," wrote Aristotle. "Even bodies on Earth do that, although they are not so great in bulk or moving at so high a speed."[20]

In Aristotle's system, matter was made up of five elements: earth, water, fire, air, and ether. Each had its own natural level—earth tended to sink down towards the center with water floating on top, air above that, and fire pushing upwards. The celestial bodies were made of the fifth element, ether. His version of the geocentric model featured some fifty-five concentric spheres. While Eudoxus may only have used the spheres as a mathematical construct (his position is not clear one way or the other),

Aristotle thought they were actual crystalline objects made of ether, all rotating around the Earth like an elaborate mobile constructed by an ingenious sculptor. The innermost body was the Moon, which he considered imperfect because it did not produce its own light. It therefore marked the transition between the corporeal world and the heavens.

Around 150 AD, Ptolemy of Alexandria came up with a new version of the model which provided an even better fit to the available data. It was known that the brightness of the planets, and the size of the Moon, varied slightly over the course of their orbits, suggesting that their distance from the Earth changed with time. Rather than ditch the circle hypothesis—as Ptolemy noted, circular motions alone are perfectly regular and therefore "strangers to disparities and disorders"—he just arranged them in a different way.[21] Each planet moved in an epicycle, which was a small circle with its center on another larger circle. The circles rolled around in such a way that the planet would produce the correct retrogradations, to an accuracy of about the size of a full moon.

Following Pythagoras and other Greek astronomers, Ptolemy believed that the cosmos was based on musical harmony (Kepler later wrote that "he seems to have recited a kind of Pythagorean dream rather than advancing philosophy"[22]). In his work *Harmonics* Ptolemy proposed exact tones for each of the heavenly bodies, which in modern tuning were equivalent to the following:

Fixed stars	D
Saturn	C
Jupiter	G
Mars	F
Sun	D
Venus/Mercury	C
Moon	G

He was so pleased with this system that he had it engraved on a stone slab outside of Alexandria.

Symmetry and Proportion

Ptolemy's model was accurate enough that it could be used as a serious predictive tool, and it served as a potent demonstration of the power of mathematics. It remained the standard model for cosmology in the

West for some twelve centuries until it was finally overturned in the Renaissance along with Aristotelian physics.

The emphasis on symmetry and harmony also influenced other fields such as art and architecture. If the cosmos was based on harmony, then it followed that human works should make use of the same patterns. In his book *De architectura*, or *Ten Books on Architecture*, the Roman architect Vitruvius (c. 80 BC–15 BC) argued that the dimensions of rooms should be based, like musical harmony, on whole-number ratios. "Without symmetry and proportion," he wrote, "there can be no principles in the design of any temple."[23] Greek forums were traditionally square, while (according to Vitruvius) temples should either be circular or have a length/width ratio of 2:1.

Architects, wrote Vitruvius, should familiarize themselves with mathematics and music so as to better exploit these harmonies. He described a method used to amplify the actors' voices in Greek theaters, in which bronze vessels were placed "in niches under the seats in accordance with the musical intervals on mathematical principles. These vessels are arranged with a view to musical concords or harmony, and apportioned in the compass of the fourth, the fifth, and the octave, and so on up to the double octave, in such a way that when the voice of an actor falls in unison with any of them the power is increased, and it reaches the ears of the audience with greater clearness and sweetness."[24]

Like the sirens of Delphi, or the planets in the sky, actors sang the tones encoded in the tetractys. Vitruvius's work served almost as a construction manual for the Romans, and its influence can be seen in many buildings including the Pantheon in Rome, with its geometrical proportions and motifs of circles and squares.

Enduring Beauty

While medieval Indian, Islamic, Persian, and European societies made great progress in mathematics—for example with the invention of Arabic numerals and decimal fractions—the geocentric model remained all but unchallenged for centuries. Its endurance was due to a number of factors. One was that it seemed in tune with intuition: after all, there was little to indicate that the Earth was itself spinning through space.

Birds were not left behind when they flew off the ground. Also, if the Earth were moving, then one would expect the relative positions of stars to change with time, just as the set of a theater looks different depending on your seat (in fact, the distances involved are so huge that the effect is small; it was not even detected until 1838).[25]

Another reason was the model's predictive accuracy—or, perhaps more relevantly, the lack of another model which could make better predictions. It was relied on, for example, by astrologers who wanted to know the arrangement of the heavens on a particular date. Ptolemy's *Tetrabiblos* long remained the most authoritative treatise on this subject.

The rediscovery of Aristotle by medieval scholars also played a role. After the collapse of the Greco-Roman Empire, Aristotle's books and lecture notes had been preserved at the library in Alexandria (named for his pupil, Alexander the Great). In the twelfth and thirteenth centuries, his works were translated into Latin in Christian Europe. Aristotle's philosophy offered what appeared to be a complete and consistent world view, and formed the basis of the curriculum at the new universities established in cities such as Bologna, Paris, and Oxford. Degrees were structured on the *quadrivium*, devised by the Pythagoreans, of arithmetic, geometry, astronomy, and music. Aristotle became known simply as The Philosopher, and his word taken as gospel truth. In *The Banquet* (c. 1304–7), Dante wrote that Aristotle "proved that this World, the Earth, is of itself stable and fixed to all eternity. And his reasons, which Aristotle states in order to break those other opinions and to affirm the truth, it is not my intention here to narrate; therefore, let it be enough for those to whom I speak, to know, upon his great authority, that this Earth is fixed, and does not revolve, and that it, with the sea, is the centre of the Heavens."[26]

Finally, the model was adopted in modified form by the Christian Church, who saw it as consistent with the idea that God's most perfect creation (that would be us) should be located at the center of the universe. Philosopher/theologians such as Thomas Aquinas (1225–74) used it to fix humankind, the natural world, and the cosmos into a unified, stable, and rational structure whose rules were dictated by God and comprehended through reason. As Aquinas put it, "Reason in man is rather like God in the world."[27]

Aquinas proposed three attributes that make an object beautiful: *integritas, consonantia, claritas*—or, as James Joyce translated them, wholeness, harmony, and radiance.[28] The geocentric model, with its celestial bodies rotating in perfect crystalline spheres around the Earth, certainly satisfied these criteria. It was more than a model—it was an enduring work of art.

Rebirth

The Renaissance saw a resurgence of interest in ancient Greek and Roman culture. Vitruvius's work was popularized by the polymath Leon Battista Alberti (1404–72) in his own *Ten Books on Architecture*. Like Vitruvius, Alberti was something of a Pythagorean: "I am everyday more and more convinced of the truth of Pythagoras' saying, that nature is sure to act consistently, and with a constant analogy in all her operations: from whence I conclude that the same numbers, by which the agreement of sounds affects our ears with delight, are the very same which please our eyes and our mind. We shall therefore borrow all our rules for the finishing our proportions, from the musicians, who are the greatest masters in this sort of numbers, and from those particular things wherein nature shows herself most excellent and complete."[29]

Alberti defined beauty in a straightforward way as a harmony of parts, in which any change would be for the worse. He proposed that room dimensions should be based, like musical harmony, on strict mathematical ratios. Dividing possible room areas into short, medium, and long, he said that short areas should have proportions of 1:1 (square), 2:3, or 3:4. The last two corresponded in musical terms to the fourth and fifth harmonic respectively. Medium rooms could have proportions 1:2 or 4:9, described as "a double proportion plus one tone more," while long rooms could be 1:3, 3:8, or 1:4. The dimensions of rooms were therefore limited to a finite set of harmonic ratios, in the same way that the frets of a guitar limit the available tones.

The Italian architect Andrea Palladio (1508–80), considered the most influential architect of the Renaissance, wrote a relatively streamlined work called the *Four Books of Architecture*. In it, he argued that certain ratios—especially the Pythagorean ratios of 1:2, 2:3, and 3:4—produce in people a spontaneous delight "without it being known why, save by those who study to know the reasons of things."[30] (Of course, this can sometimes be inverted, such as when designers or scientists select certain patterns based on unconsciously absorbed aesthetic criteria.)

Palladio selected seven room shapes that contain the "most beautiful and harmonious proportions." These included the circle, the square, the harmonic ratios 1:2, 2:3, 3:4, and 3:5, and the ratio 1:1.414 . . . (etc.). This last number is the ratio of the side of a square to its diagonal, which equals the square root of two. The Pythagoreans had proven that this number was irrational—in other words, that it could not be represented as the ratio of two whole numbers.

Palladian architecture was exported to England by Inigo Jones after a visit to Italy, and it soon spread to other countries. As a result, many churches, museums, and monuments in Europe and North America were built according to Pythagorean proportions.

The emphasis on mathematical proportion was equally strong in art, where the new theory of perspective required a solid knowledge of geometry. Renaissance artists such as Piero della Francesca, Albrecht Dürer, and Leonardo da Vinci were accomplished applied mathematicians. Just as Plato inscribed "Let no one enter who is lacking in geometry" over the entrance to his academy, so Leonardo wrote in his notes, "Let no one read me who is not a mathematician."

An example of the close relationship between mathematics and art was the book *De Divina Proportione*, written by Luca Paciolo with illustrations by Leonardo and published in Venice in 1509.[31] It included chapters on geometry, perspective, Vitruvian architecture, and the five regular solids. The regular solids consist of the pyramid, cube, octahedron, dodecahedron (shown in Figure 1.2), and icosahedron. These highly symmetric figures have identical faces and can be inscribed in a sphere. It was proved by Euclid that there are only five such figures; no others exist. The mathematician Hermann Weyl called this "one of the most beautiful and singular discoveries made in the whole of mathematics."[32]

In *Timaeus*, Plato proposed that these solids were the basis for the four elements: earth (cube), air (octahedron), water (icosahedron), and fire (tetrahedron). He associated the dodecahedron with the universe as a whole, since it was the closest to a perfect sphere. As physicist John D. Barrow notes, these solids "remain objects of aesthetic appeal and geometric fascination to mathematicians. The models that are made of them never cease to amaze us with their combination of beauty, symmetry, and simplicity."[33]

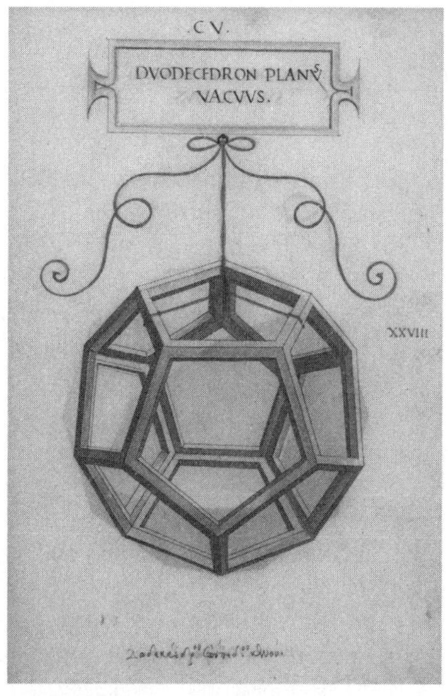

Figure 1.2 A dodecahedron from *De Divina Proportione*, written by Luca Paciolo, illustration by Leonardo Da Vinci. By showing only the edges and using perspective, Leonardo's illustration allows us to visualize the three-dimensional shape. Source: Reprinted from Luca Pacioli, *Divina proportione opera a tutti glingegni perspicaci e curiosi necessaria* (Venice: A. Paganius Paganinus, 1509).

While Leonardo was fascinated by mathematics and symmetry, he was also one of the first artists to base his drawings on a close study of nature, including anatomical studies of human beings (see Box 1.2). It was a similar fusion between theory, observation, and practice which inspired Renaissance scientists to finally produce an alternative model of the universe.

Divine Revolutions

Nicolaus Copernicus wasn't the first person to suggest that the Earth might rotate around the Sun. The Greek astronomer Aristarchus (310 BC–ca. 230 BC) had said something similar nearly two millennia earlier; however, the opinions of Aristotle held sway.

Copernicus's *On the Revolution of Heavenly Spheres* was published in 1543, while he was on his deathbed. The Polish astronomer had been working on his theory for three decades, since he was a student in Italy, but had delayed its publication. With cosmology being such a central part of Church dogma, the idea that the Earth was not the navel of the universe seemed bound to receive a hostile reception.[34]

Box 1.2

Vitruvian Man

The Roman architect Vitruvius believed that even the human body was based on geometrical principles. He claimed that "in the human body the central point is naturally the navel. For if a man be placed flat on his back, with his hands and feet extended, and a pair of compasses centered at his navel, the fingers and toes of his two hands and feet will touch the circumference of a circle described therefrom. And just as the human body yields a circular outline, so too a square figure may be found from it. For if we measure the distance from the soles of the feet to the top of the head, and then apply that measure to the outstretched arms, the breadth will be found to be the same as the height, as in the case of plane surfaces which are perfectly square."[35]

Unfortunately, when illustrators later followed his instructions for the human figure literally, the results looked rather ill-proportioned, as shown by the fifteenth-century woodcut in Figure 1.3. The arms are pulled out in order to meet the demands of abstract symmetry, as if the subject is being stretched on the rack. The hands and feet are also too large.

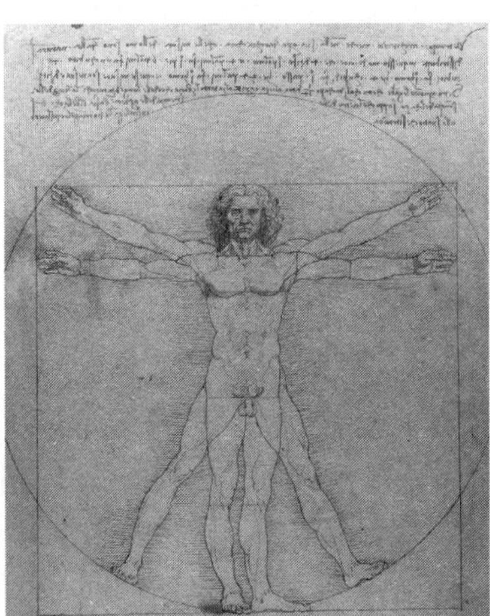

Figure 1.3 Left panel is an illustration from a 1521 translation by Cesare Cesariano of Vitruvius's *Ten Books on Architecture*. Right panel is Vitruvian Man (c. 1487) by Leonardo da Vinci. The mirror writing is the text by Vitruvius describing the human figure in geometric terms. Source (left): Reprinted from Vitruvius, *De Architectura*, trans. Cesare Cesariano (Como, Italy: Gottardo da Ponte for Agostino Gallo and Aloisio Pirovano, 1521). Source (right): Leonardo da Vinci, *Vitruvian Man*, 1487. Galleria dell' Accademia, Venice.

Leonardo's version of Vitruvian Man relaxed the demands of mathematical symmetry. It combined two different views based on the circle and the square, with the circle centered on the navel and the square centered slightly lower. The result is less exactly aligned in mathematical terms, but is a much better representation of the human figure and has become an iconic image.[36]

In a preface, Copernicus wrote that he had considered following the example of the highly secretive Pythagoreans, who refused to disseminate their results on the basis that "they wanted the very beautiful thoughts attained by great men of deep devotion not to be ridiculed."[37] However, his associates had talked him into going ahead. The preface was later cut by the editor and replaced with a polite letter saying that the model was not intended to be a representation of the truth, but was only a kind of mathematical trick that made calculations easier.

In fact, the religious response to the book was relatively muted, probably because it was not taken seriously by most astronomers. The new system was not as good at making predictions as the standard Earth-centered model proposed by Ptolemy. Nor did it completely get rid of epicycles. Its main appeal, as Copernicus noted in his text, was aesthetic: it provided a more elegant, symmetric, and harmonious solution to a number of problems.

In the introduction, Copernicus argued that astronomy is concerned with beauty: "Among the many various literary and artistic pursuits which invigorate men's minds, the strongest affection and utmost zeal should, I think, promote the studies concerned with the most beautiful objects, most deserving to be known. This is the nature of the discipline which deals with the universe's divine revolutions . . . and which, in short, explains its whole appearance."

The Sun-centered model gave a simple and elegant explanation for why the size and brightness of planets appeared to change, since their distance from us would vary depending on their position in their respective orbits. And retrograde motion could be explained in terms of the combined circular motions of the planets and the Earth, which move at different speeds. "In this arrangement, therefore," wrote Copernicus, "we discover a marvelous symmetry of the universe, and an established harmonious linkage between the motion of the spheres and their size, such as can be found in no other way."

Putting the Sun at the center was of course a radical change, but "in this most beautiful temple, who would place this lamp in another or better position than that from which it can light up the whole thing at the same time? For, the sun is not inappropriately called by some people the lantern of the universe, its mind by others, and its ruler by still others."

Copernicus maintained that the celestial bodies and the universe as a whole are spherical, because "of all forms, the sphere is the most perfect. . . . Hence no one will question the attribution of this form to the divine bodies." Similarly, "the motion of the heavenly bodies is circular, since the motion appropriate to a sphere is rotation in a circle."[38] This insistence on circular motion meant that, despite its aesthetic appeal, the new model still required epicycles to capture subtle effects. More importantly for people like astrologers, the model lacked predictive accuracy (Copernicus preferred theorizing to obtaining data; he rarely made astronomical observations himself).

Before the Copernican model could prove its worth, therefore, it had to go even further than shifting the Earth from the center of the universe—it had to get rid of the circles. That task would be left to the German astronomer/astrologer Johannes Kepler.

Harmony of the World

On 9 July 1595 (he later recorded the date), while drawing a geometric figure on a blackboard during a class he was teaching, Kepler was struck by a sudden flash of insight. For years, he had been working with the Copernican model and pondering its structure. Why were there six planets and not a hundred—or a thousand? What explained their distances from the Sun and rates of rotation? Was there a deeper pattern? Kepler was deeply religious and believed that God had created an ordered universe. If the Copernican system was part of his plan, then there had to be some reason behind it.

His epiphany was that the arrangement of the cosmos could be explained by geometry. The figure he had drawn for his school class, for unrelated reasons, was a triangle inscribed between two circles. Looking at the circles, he noticed that their relative sizes, which differ in radius by a factor of two, were (to within the margin of observational error) the same as those of the orbits of Saturn and Jupiter. Perhaps the orbits of the other planets could similarly be explained by inscribing geometric figures between them—in which case the Copernican system would be revealed as a stunning demonstration of divine order.

Figure 1.4 The figure which inspired Kepler. The diameters of the inner and outer circles differ by a factor of two. For comparison, the ratio of the average distances of Saturn and Jupiter from the Sun is about 1.83. Source: Author

Kepler first tried different arrangements of 2D polygons, but soon realized that it made more sense to use the five regular solids. These gave a much better fit and also explained the number of planets; for if there were only five regular solids to fit between the orbits, then there could only be six planets. This model, shown in Figure 1.5, was so beautiful that he knew it had to be true. As he later wrote, "Within a few days everything fell into place. I saw one symmetrical solid after the other fit in so precisely between the appropriate orbits, that if a peasant were to ask you on what kind of hook the heavens are fastened so that they don't fall down, it will be easy for thee to answer him."[39] He even wondered if the secret was known to the ancients: "What if the Pythagoreans did teach the same thing that I do, weaving their meaning into a cover of words?" Kepler tried to sell a working silver version of the model to the Duke of Württemberg, but the project fell through because of complication and expense.

While the model seemed a good fit, the quality of the available data in the age before telescopes wasn't good enough to verify the theory with complete, 100-percent accuracy. Earlier scientists would probably have declared victory, pronounced the theory true, and hit the lecture circuit, but Kepler decided to seek confirmation using the best available

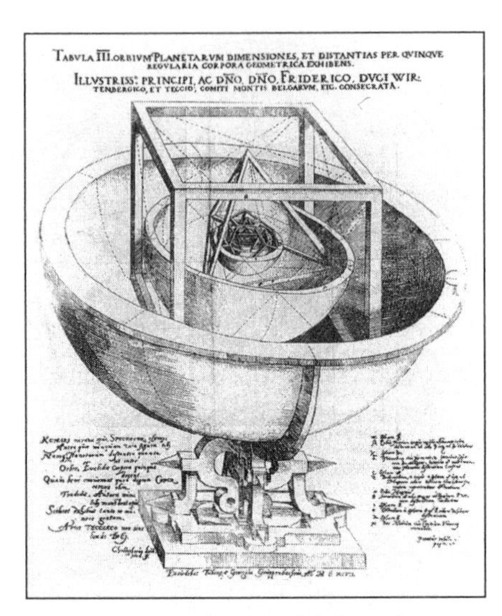

Figure 1.5 Kepler's model, based on the five regular solids, from *Mysterium Cosmographicum*. Note the use of Leonardo's style of drawing with open-faced solids, as in Figure 1.2. Source: Johannes Kepler, "Solar System," in *Mysterium Cosmographicum* (Tübingen, Germany: Munchen, Munchener druck- und Verlagshaus, 1596).

data. Unfortunately for him, that was in the hands of a cantankerous Danish nobleman named Tycho Brahe, who had the finest (naked-eye) observatory in the world but was jealously possessive of anything that came out of it, and furthermore had his own theories which he wanted to prove.

Kepler would eventually obtain his data, but only—to cut a long story short—by taking a job with Brahe, becoming his tenant, and inheriting the data after Brahe died prematurely, probably from mercury poisoning. When at last Kepler could compare the data with his model, he found to his chagrin that there were significant discrepancies. No matter how hard he tried, he couldn't get his circles to perfectly match.

After years of labor, he finally discovered—or rather was forced to admit—that the orbits were not circles at all, but were ellipses, with one of the two foci being the Sun. This observation constituted the first of his three laws of planetary motion:

1. The planets rotate around the Sun in elliptical orbits.
2. Orbiting planets sweep out equal areas in equal time.
3. The squares of the period are proportional to cubes of average distance from the Sun.

While these three laws expressed patterns and a kind of stability in the data, Kepler couldn't explain why the planets behaved in this way—and without the symmetry and beauty of perfect circles his ellipses lacked appeal, even to him.[40] In his 1619 book *Harmony of the World* he therefore preferred to concentrate on his theories of mathematical harmony as applied to astrology, meteorology, and astronomy. "Geometry," he wrote, "is the archetype of the beauty of the world." Following the Greeks, he worked out the notes played by the celestial orchestra, but included the effects due to their variations in angular velocity as they swept around the Sun. He concluded that "the Earth sings Mi-Fa-Mi, so we can gather even from this that Misery and Famine reign on our habitat."[41] (Kepler's life was not easy: the countries he lived in were constantly at war, employment was unstable, several of his children died from illnesses, his mother was tried for witchcraft, and so on—so ellipses were really the least of his problems.)

Other scientists also preferred to stay with circles. Even when Galileo published his *Dialogue Concerning the Two Chief World Systems* (1632), which argued for the Copernican model and brought him into conflict with the Catholic Church, he ignored Kepler's version. That reluctance to give up the ancient symmetries would only disappear when Isaac Newton finally revealed that within the Copernican system lurked a more profound type of circular law than anyone had foreseen.

The Clockwork Universe

In a 1605 letter, Kepler wrote that "my aim is to show that the machine of the universe is not similar to a divine animated being, but similar to a clock."[42] While Kepler couldn't say what made the clock tick—he suspected magnetism—Newton showed that it was the force of gravity. As Newton proved in his *Principia Mathematica* (1687), Kepler's laws of planetary motion were just the solutions to Newton's equations when applied to orbiting bodies.

The story goes that the law of gravity came to Newton when he was sitting in his garden and saw an apple fall to the ground. Scientific discoveries often seem to come in the form of a sudden epiphany, when the solution is presented up as a complete vision to the conscious mind. Newton's discovery, of course, would prove more useful than Kepler's original insight that the universe was based on regular solids, but again it was inspired by harmony and symmetry: the force which makes an apple fall to the ground and that which embraces the Moon in its orbit are one and the same.

In one sense, the law of gravity vindicated the ancient belief that the universe is based on circles and spheres. Indeed, as Alfred North Whitehead observed, "seventeenth century science reads as though it were some vivid dream of Plato or Pythagoras."[43] Newton's law of gravity has a spherical symmetry in the sense that gravitational force depends only on distance—so, for example, any two bodies that are the same distance from the Earth's center will experience the same gravitational pull. The heavenly bodies are round, not because spheres are the only allowable shape as Copernicus had said, but because the force of gravity that holds them together is spherically symmetric. The symmetry therefore shifts from the phenomenon—the actual observed shapes or orbits—to the mathematical law.

"The *Principia*," wrote author and literary critic Marjorie Hope Nicolson in 1959, "offered climactic proof that order, proportion, regularity were universal principles, comprising the harmony of the universe. More than ever before, man could turn to the cosmos for his ethics of limitation, his aesthetics of order and proportion."[44] As seen later, the search for mathematical harmonies and symmetries still motivates and defines much of our modern scientific program, especially in areas such as supersymmetric string theory.

If nature is governed by physical laws, then it follows, as Newton wrote in *Principia*, that we can understand it "by the same kind of reasoning as for mechanical principles." The "historical goal" of physicists, according to Steven Weinberg, "is the formulation of a few simple principles that explain why everything is the way it is. This is Newton's dream, and it is our dream."[45] In the next chapter, we will see how Newton's "rational mechanics" was used to build a model of model of the cosmos from the bottom up.

2

Integrity

The chief forms of beauty are order
and symmetry and definiteness, which the
mathematical sciences demonstrate
in a special degree.

Aristotle, *Metaphysics*

The whole object's form is its perfection
and arises out of the integrity of its parts.

Thomas Aquinas, *Summa Theologica*

The beauty that Nature has revealed to
physicists in Her laws is a beauty of design,
a beauty that recalls, to some extent, the
beauty of classical architecture, with its
emphasis on geometry and symmetry.

Anthony Zee, *Fearful Symmetry*

The first step in reductionist science is to reduce a system to its components. For this to work there has to be a smallest unit to reduce to—otherwise the process will go on without end. For the Pythagoreans, the building blocks of the universe were numbers, or more specifically the positive integers (from the Latin word integer *for whole or complete). For Aristotle and his followers, they were the five elements, which were associated with the regular solids. For the Greek philosopher Democritus, they were atoms. The eighteenth and nineteenth centuries saw the discovery of the chemical elements and the first subatomic particle. This chapter shows how aesthetic principles served as both a guide and an inspiration in the search for integrity and solidity in a sometimes ephemeral world.*

Perhaps the most famous theory of science, and certainly one of the oldest, is the notion that matter is made up of discrete, indestructible particles. This theory was developed and championed in the fourth century BC by Democritus, who named the particles *atomos*, or atoms, from the Greek for indivisible. While the theory would eventually be accepted by scientists—today it is possible to sense and manipulate individual atoms—its original motivation had as much to do with philosophy and aesthetics as it did with science.

As a young man, Democritus inherited a large sum of money from his father. He promptly blew it all on foreign travel. When he returned, according to his biographer Diogenes Laertius, he lived in "a most humble manner; like a man who had spent all his property, and that on account of his poverty, he was supported by his brother Damasus."[1] However, he soon became famous for his ability to make accurate predictions of natural phenomena such as storms or earthquakes, and "was for all the rest of his life thought worthy of almost divine honors by the generality of people." He became a pupil of Leucippus, who is believed to have played a large role in the atomic theory, though little is known of his work.

Democritus published over fifty books on a huge variety of topics, including music, medicine, and geography. During his travels, it seems that he was exposed, perhaps through Philolaus, to the ideas of Pythagoras. Laertius wrote that "he appears to have derived all his doctrines from him to such a degree, that one would have thought that he had been his pupil, if the difference of time did not prevent it."

Democritus believed that nothing existed in the universe except for atoms, which were eternal and immutable, and the void, or empty space. Like the Pythagorean tetractys, matter was made up of an arrangement of dots. The qualities of a substance reflected the nature of its atoms, which differed only in shape and size. In a heavy material such as lead, the atoms were closely spaced, while in a light material they were more spread out. Oil was made of small, round atoms that slid over one another, while the atoms of rocks had small hooks that held them strongly in position. Fire atoms were barbed and painful to the touch. A substance's large-scale properties were therefore a direct reflection of what was happening at the atomic level. As Aquinas later wrote, the object's form "arises out of the integrity of its parts."[2]

According to one story, the atomic theory was inspired by the smell of fresh bread. How, Democritus asked, is it possible to detect the presence of a loaf of bread even before it is brought into the room? The answer must be that the particles of the bread—the atoms—waft ahead of it and interact with the atoms of our nostrils. Subjective sensations like taste and smell were nothing but the interplay between the atoms of our bodies with those of the environment: "Sweet exists by convention, bitter by convention, color by convention; but in reality atoms and the void alone exist."[3]

The theory also provided an explanation for an even more basic phenomenon than the smell of bread—namely, how it is possible for things to move from one point to another. As Zeno pointed out, this led to a logical conundrum. Suppose that you want to walk across the room. First, you have to travel half the distance. Then you have to travel half of the remaining distance (a quarter of the way), and half of the remaining distance (an eighth), and so on. Since this series goes on forever, Zeno argued that distance must be unlimited, and that you can never arrive at your destination. Since people evidently did manage to cross the room in a finite amount of time, there appeared to be a contradiction. The atomic theory offered a way out: it proposed that distance had a fundamental lower limit so that the series didn't go on forever (see the notes for another solution).[4]

The real appeal of the theory, however, had as much to do with the way it imposed a calming kind of order upon the chaos of the world. Like a painting of a beautiful sunset, or a pleasant wallpaper, it gave a reassuring sense that everything was in its place.

The Aesthetics of Calm

The aim of art is not always disruption or revolution—it is also an effective way to comfort and console, as any interior decorator knows. Pythagoras, for example, was said to use music as a form of art therapy. According to Iamblichus, late one night Pythagoras was walking home after doing some astronomy when he came across a young man who was incensed that his girlfriend was with another lover. Before the man headed off to burn down her house, Pythagoras ordered him to listen to some soothing music (today he might have played recordings of whale songs or the rainforest). The man immediately cooled down and "returned home in an orderly manner."[5]

The atomic theory was similarly promoted as a kind of calming device for the Greek soul. Its attraction was not so much its advancement of knowledge—after all, no one could see these hypothetical atoms—but *ataraxia* or peace of mind. Democritus praised detachment and the ability to be astonished at nothing, and atoms provided a reassuringly mechanistic explanation for the world.[6] The atomic theory was soon adopted as part of a wider campaign against irrational religious beliefs and in favor of materialism. "The reason why all mortals are so gripped by fear," wrote Lucretius in his philosophical poem *The Nature of the Universe*, "is that they see all sorts of things happening on the Earth and in the sky with no discernible cause, and these they attribute to the will of the gods."[7]

In the atomic universe, everything we see around us—even our own bodies—is nothing but a transient configuration of atoms. Matter is never created or destroyed, only rearranged. The world is a vortex of continuous change, but the underlying constituents are eternal and unchangeable. There is no free will, because everything is determined by the flow of atoms: as Democritus said, "Everything existing in the universe is the fruit of chance and necessity." Even death should hold no fear, since it only represents dissolution into our original components, which themselves are eternal.

Democritus was known as the laughing philosopher because of his emphasis on cheerfulness (the atomistic calming therapy apparently worked for him, at least). According to Laertius, he lived to the ripe old age of 109. When he was near death, his sister was concerned that he would die during an important festival. He asked that hot loaves of bread be brought to him every day. He did not eat, but only smelt the bread—he inhaled the atoms. When the three-day festival came to a close, he died without pain.

Sense of Humor

Atoms may last forever, but the atomic theory faded away soon after Democritus, before eventually wafting back into style. The theory had two major flaws. The first was that it was hard to prove, atoms being very small and therefore impossible to observe directly. The second problem, which was even more serious, was that Aristotle didn't like it.

Aristotle had his own institution, the Lyceum, along with many students, and so was in a position of considerable power to propagate his opinions—unlike Democritus, who complained that "I came to Athens, and no one knew me." He was opposed to the idea of the void, which as Parmenides had pointed out seemed to lead to contradiction. For if we suppose there is a void, it then follows that the void is not nothing, hence it is not really a void (such logical conundrums caused the Greeks a great deal of trouble).

Aristotle instead agreed with earlier philosophers like Empedocles that matter was made up of the four elements of earth, water, air, and fire. This theory was actually a composite of four previous versions: Thales, who is sometimes known as the first philosopher, believed that everything was made of water; Anaximenes that it was made of air; Heraclitus fire; and Xenophanes earth. In his *Doctrine of the Four Elements*, Empedocles argued that the elements combine in fixed Pythagorean proportions. Bone, for example, consisted of four parts fire, two parts water, and two parts earth.

According to Aristotle, these elements varied in a continuous way so that there was no smallest atom and no void. They could also transform into another, so water could evaporate to form air or be dried to form earth. Physicians including Hippocrates linked the elements to the four bodily "humors": black bile (earth), phlegm (water), blood (air), and yellow bile or pus (fire). In a healthy person the humors were in balance with one another, while an imbalance could cause disease. For example, an excess of black bile was associated with cancer.

Because Aristotle's influence extended well into the late sixteenth century, the idea of the four elements persisted along with the geocentric model. Alchemists were fascinated by the idea of transmuting substances into new ones, especially if the end result was gold. Physicians employed dubious medical treatments, such as bleeding or forced vomiting, to restore equilibrium of the bodily humors. A person's temperament was also determined by the composition of these humors, which could make a person bilious, phlegmatic, sanguine (from the Latin for blood), or jaundiced (from the Latin for yellow). A Shakespearean curse (from *Pericles*) was:

> Thou hast as chiding a nativity
> As fire, air, water, earth, and heaven can make,
> To herald thee from the womb[8]

Shakespeare himself acted in the play *Every Man in His Humour* (1598) by Ben Jonson, in which characters represented particular humors.

Although the Aristotelian model was still the official church orthodoxy in Shakespeare's time, astronomers were beginning to take issue with some of its ideas. Aristotle had taught that the Earth is still, but in 1543 came Copernicus's unsettling proposal that it was actually in very fast motion around the Sun. Aristotle's model of the universe was slowly being replaced by one based on mathematical theory backed by empirical observations. And once again, like Democritus, scientists were waking up to the smell of freshly baked bread and wondering whether the world might not be made of atoms and space.

Lack of Integrity

Galileo was open to the idea of atoms—he mentions them in *Two New Sciences* and other works—and Newton thought that all matter, and even light, was made up of "solid, massy, hard, impenetrable, movable particles." But because of the inconvenient smallness of atoms, they remained a topic for philosophical debate until the late eighteenth century, when new empirical evidence for their existence began to emerge.

The French chemist Antoine Lavoisier did not himself endorse the atomic theory, but he played a key role in proving its veracity. Trained as a lawyer, he made enough money as part-owner of a private tax collection firm to establish a private laboratory, with his wife Marie Anne Lavoisier as his assistant. His experiments, which are considered amongst the most beautiful in science, applied an accounting-like rigor to chemistry and changed the way we think about earth, water, air, and fire.

Earth. Aristotle had argued that one element could transform into another, and even in 1770 many believed that water could be transformed into earth if it was heated long enough—for when water was heated over a period of days in a glass container, it left a thin sediment. Lavoisier tested this by boiling water for a hundred days in a special flask in which the water vapor condensed and was returned to the flask, so that no water could leave the system. By carefully comparing measurements of both the water and the flask before and after the experiment, he showed that the water did not change its weight, so it was not being transformed. However, the flask weighed slightly less. The difference was made up by the sediment, which consisted not of water or earth, but of material from the glass. Lavoisier thus proved that earth was not being produced.

Fire. Another phenomenon which had long troubled both scientists and alchemists was the nature of fire. The dominant theory was that when a material burned, it released a substance called phlogiston (from the Greek word *phlogistos* for flammable) which corresponded to Aristotle's element of fire. Because nothing burned without air, it was assumed that air was an ingredient of the phlogiston. In 1753, Robert Boyle found that when a metal burned, its weight did not decrease as expected from a loss of phlogiston, but rather increased. This appeared to disprove the theory, though some simply argued that phlogiston had a negative weight, or as the Marquis de Condorcet put it was "impelled by forces that give it a direction contrary to that of gravity."[9] The matter was not resolved until Lavoisier—again using careful measurements—showed that when phosphorus and sulfur are burned in a fixed volume of air, the weight gained came from the air.

Air. Lavoisier showed that air consisted of a mixture of gases, primarily oxygen (his name for it) and nitrogen. Collaborating with the mathematician Pierre-Simon Laplace, he also demonstrated that the processes of respiration and digestion are analogous to the slow burning of food in the presence of oxygen. The experiments involved placing a guinea pig in a cold container surrounded by ice to measure how much ice was melted from the heat generated by its body. This gave us the phrase "guinea pig" for the unwilling subjects of scientific experiments.

Water. In a two-day public demonstration in 1785, Lavoisier proved that water was made up of two substances: hydrogen and oxygen. He first decomposed water into hydrogen and oxygen by passing water vapor over incandescent iron. The two gases were collected, mixed together in a glass vessel, and ignited by an electric spark to form water vapor.

With the help of his wife, Lavoisier summed up his life's work and laid down the foundations of modern chemistry in his book *The Elements of Chemistry*.[10] This included the first explicit statement of the conservation of mass. Empedocles wrote that "it is impossible for anything to come to be from what is not," but Lavoisier demonstrated the principle by use of the chemical balance. His book introduced a rigorous system of chemical nomenclature and proposed that substances are of two types: elements and compounds. These elements—unlike the ancient version—cannot be broken down, while compounds like water are made up of one or more elements. He listed thirty-three known elements, with the disclaimer that future experiments might result in revisions. Eight did later turn out to be composites, and his list also contained heat and light, which fell under the category of "imponderable fluids."

The Pythagoreans, who believed that the universe was made up from numbers, had associated odd numbers with the divine and even numbers with duality and conflict, because they could be divided in two. By this standard, showing that Aristotelian elements like air and water were divisible was like calling them weak and ugly. Instead of being solid and self-contained, or representing anything fundamental, they were just a provisional conglomerate of parts that could be broken to pieces at any time. They lacked the aesthetic quality that Aquinas called *integrita*, or integrity.

Lavoisier's book was published in 1789, the same year as the storming of the Bastille. Being a wealthy nobleman and part-owner of an unpopular taxation firm, Lavoisier soon found himself a target of the

Revolution. On 8 May 1794, he was taken to the guillotine and executed. The mathematician Joseph-Louis Lagrange remarked, "It took only a moment to cause this head to fall, and a hundred years will not suffice to produce its like."[11]

A century later, a statue of Lavoisier was put up in Paris—but in a final indignity it later emerged that the artist had based the head on the Marquis de Condorcet, the champion of phlogiston. During the Second World War, the statue—which also lacked integrity—was melted down and never replaced.

Whole Numbers

One of the most appealing features of the Aristotelian model, from an aesthetic point of view, had been its simplicity. In this sense, Lavoisier's work was a step in the wrong direction. Instead of Aristotle's four elements, the list now contained things like oxygen, hydrogen, nitrogen, mercury, phosphorus, sulfur, and zinc. The world was getting more complex, rather than simpler. If there was such a thing as an atom, it had to exist in many different forms. This did not deter the English scientist John Dalton, who resurrected the atomic theory by finding a different kind of simplicity and integrity.

In the early 1800s, it was becoming clear that when elements combined to form compounds, they always did so in the same ratio by weight. Furthermore, just as Pythagoras had found with musical harmony, the proportions always seemed to involve whole numbers. For example, when hydrogen and oxygen were burned to form water, as in Lavoisier's experiment, one gram of hydrogen always combined with eight grams of oxygen—so the ratio was 8:1, not something like 8.2:1. As Dalton realized, this Pythagorean exactness could be explained by the existence of atoms. When substances combined, the individual atoms of each substance were fused in a fixed ratio.

Dalton's interest in atoms was inspired by his study of weather. From 24 March 1787 until 26 July 1844, the day before his death, he compiled a detailed list of meteorological observations, and his first major published work was *Meteorological Observations and Essays.*[12] In an appendix, he discusses the composition of the atmosphere in terms of intermingling "particles" of water vapor and air. He believed that the pressure of gases was caused by the force of repulsion between individual particles. His atomistic approach also extended to language; in 1801 he wrote a manual called *Elements of English Grammar*.[13]

In 1808, Dalton published *A New System of Chemical Philosophy*, which described his atomic theory. Echoing Democritus, he wrote, "Matter, though divisible in an *extreme* degree, is nevertheless not *infinitely* divisible. That is, there must be some point beyond which we cannot go in the division of matter. . . . I have chosen the word atom to signify these ultimate particles."[14] Like Democritus, he believed that atoms were immortal and unchanging: "No new creation or destruction of matter is within the reach of chemical agency. We might as well attempt to introduce a new planet into the solar system, or to annihilate one already in existence, as to create or destroy a particle of hydrogen." Atoms of a particular element were always identical, but they could combine with other elements, which had distinct atoms, to form different compounds. All change consisted of "separating particles that are in a state of cohesion or combination, and joining those that were previously at a distance." Matter might be in a state of perpetual flux, but atoms lived forever. The particles that made up the atmosphere were the same as the ones Democritus had breathed in over two millennia before, and they had existed since the time of creation.

Dalton's book included a list of the relative atomic weights of different elements, deduced by observing how they combined to form compounds. Hydrogen, being the lightest element, was assigned a weight of 1, while oxygen was 7, iron was 38, gold was 140, and mercury topped the list at 167. The accuracy of the list was hampered by measurement error and Dalton's assumption that elements combined in equal ratios. For example, he thought of water as equal parts hydrogen and oxygen, while in fact there are two atoms of hydrogen for every oxygen atom.

The units of atomic weight are now called Daltons in his honor. At the time, though, few scientists were ready to adopt the atomic theory. The experimental evidence was not strong enough to convince them of the existence of tiny, invisible atoms, and the theory was not without problems. Some elements turned out to have non-integer atomic weights—such as chlorine, which has an atomic weight of 35.45—which somewhat diminished the theory's Pythagorean appeal (as discussed in Chapter 5, it was later found that chlorine comes in a mix of different versions known as isotopes). The atomic theory therefore remained a kind of theoretical device until it gained fresh life from yet another emerging pattern—this time based on octaves.

The Law of Octaves

Between 1800 and 1860 the number of known elements nearly doubled from thirty-one to sixty. Scientists were also learning more about how the elements combined to form compounds. For example, an atom of oxygen combined with two atoms of hydrogen to form water; an atom of nitrogen with three atoms of hydrogen to form ammonia; and an atom of carbon with four atoms of hydrogen to form methane, or with two atoms of oxygen to give carbon dioxide.

These discoveries led to the idea of valency. Atoms were thought of as bonding together using imaginary hooks, and the valency was the number of hooks for a particular element. The valency of hydrogen was one while that of oxygen was two, so two atoms of hydrogen could combine with one atom of oxygen. Nitrogen had a valency of three, so it bonded with three hydrogens. Carbon had a valency of four, so it could bond with four hydrogens or two oxygens.

A number of scientists soon noticed that the valency of elements was related to atomic weight. Elements whose atomic weight differed by eight showed similar chemical properties, following what the English chemist John Newlands—who had a musical background—called a "law of octaves."[15] The breakthrough finally came when the Siberian chemist Dmitri Mendeleev, who was working on a chemistry textbook, laid out the sixty-three elements known at the time by order of their atomic weight and saw that there was a rhythmic pattern in their valences, going 1,2,3,4,3,2,1 and so on in repetition.[16] One night, he later said, he had a dream "where all the elements fell into place as required. Awakening, I immediately wrote it down on a piece of paper." Mendeleev was a card player, and he found that the elements could be dealt out in a rectangular grid, with a new row beginning each time the valency cycle repeated.

Elements in a particular column then had the same valency, and therefore similar chemical properties.

Blanks would sometimes appear in the table where no known element seemed to fit, and Mendeleev conjectured that these corresponded to new, undiscovered elements. Sure enough, gallium was discovered in 1875, scandium in 1879, and germanium in 1886. This was convincing proof that atomic weight had a direct, though still unclear, correspondence with chemical properties. The patterns could do more than explain—they could predict.

In 1893, as an award for his discovery, Mendeleev was appointed Director of the Russian Bureau of Weights and Measures, where his work included determining the optimal alcohol content for vodka (the answer, after much research, was 40 percent alcohol by volume, which is 80 proof). He was certainly qualified for the position, since his doctoral dissertation had the title "On the Combinations of Water with Alcohol."

By the end of the nineteenth century, Democritus's idea that matter was made up of indivisible atoms whose properties determined how substances behave was increasingly accepted. "Sweet exists by convention, bitter by convention, color by convention; but in reality atoms and the void alone exist." The plurality of substances—with all their unique textures, smells, colors, and behaviors—had been reduced to a harmonized table of atomic weights. Qualities had been replaced by quantities.

It was later found that the valency of a substance—its hooks and barbs—reflected the internal structure of the atom. But that would follow after scientists made discoveries about a seemingly unrelated topic: electricity.

Current Thinking

Around 600 BC, Thales discovered that when amber was rubbed with fur it would attract light objects such as hair or straw. If it was rubbed hard enough it could even produce a small spark. There the state of electrical engineering rested for over 2,000 years until 1600 when William Gilbert, who was personal physician to Queen Elizabeth I, published his book *De Magnete,* which summarized his investigations into electricity and the related phenomenon of magnetism. His book coined the term "electric" from the Greek *elektron*, for amber; and it contained a list of substances—including precious stones such as diamond, sapphire, and opal—that he found could be charged by rubbing. It also gave Kepler his idea that the planets could be swept around the Sun by some kind of magnetic force.

In 1734, the French physicist Charles Du Fay, whose day job was superintendent of the Jardin du Roi in Paris, published a paper showing that electricity actually existed in two forms, which he called vitreous (produced by glass) and resinous (produced by resin): "The first is that of Glass, Rock-Crystal, Precious Stones, Hair of Animals, Wool, and many other Bodies: The Second is that of Amber, Copal, Gum-Lack, Silk, Thread, Paper, and a vast Number of other Substances."[17] As with magnetic poles, like charges repelled one another while opposites attracted.

Another leading experimenter was Benjamin Franklin, who in 1752 famously flew a kite in a storm and used it to collect the charge from a lightning strike, thus conclusively proving that lightning is another form of electricity. He probably also came close to electrocuting himself—a fate suffered by some who imitated his experiment. Franklin believed that electricity was a kind of fluid. If an object had an excess of the fluid he called it positively charged, while an object with a shortage of the fluid was negatively charged. This naturally led to the idea of conservation of charge—as with mass conservation, the electricity was neither created nor destroyed, but just moved from one place to another. Somewhat arbitrarily, Franklin identified positive charge with Du Fay's vitreous electricity and negative charge with the resinous kind.

In 1785, Du Fay's compatriot Charles Coulomb proved that the electric force of attraction or repulsion decreased, like Newton's gravity, with the distance squared.[18] Once again, a fundamental force of nature was shown to obey a simple, aesthetically pleasing mathematical equation. It followed that electrical systems could be understood and controlled through mathematical principles.

During the nineteenth century, scientists used new devices like Alessandro Volta's battery to build a complete theory of electricity and magnetism. In 1820, the Danish physicist Hans Ørsted discovered that an electric current flowing in a wire produces a magnetic field around the wire.[19] (Permanent magnets, of the sort that now attach to fridges, work in a similar way because their atoms are aligned in roughly the same direction. Electrons circulating around the atomic nuclei form many small loops which together create a magnetic force. An additional contribution comes from each electron's spin around its own axis.)

The British physicist Michael Faraday found in turn that a changing magnetic field induces an electric field, showing that electricity and magnetism were profoundly related by some kind of deep symmetry: when one moved, the other reacted. This discovery prompted Faraday to express his belief in the unity of natural forces: "I have long held an opinion, almost amounting to conviction, in common I believe with many other lovers of natural knowledge, that the various forms under which the forces of matter are made manifest have one common origin."[20]

Finally, in 1865, the mathematical physicist James Clerk Maxwell tied it all together in four neat equations, which showed that electric and magnetic forces are two complementary aspects of a single phenomenon known as electromagnetism. Along with Newton's law of gravity, Maxwell's equations are an archetypal example of a beautiful scientific theory that reduces a broad range of phenomena—in this case, everything to do with electricity and magnetism—to a few short lines. When physicists such as Leon Lederman describe the hoped-for Theory of Everything as something that can fit on a T-shirt, they have in mind Maxwell's equations, which are available not just on T-shirts but also baseball caps, tote bags, thongs, throw pillows, and so on.[21]

The equations did more than show how an electric current produces a magnetic field, or how a changing magnetic field produces a current in a wire; they also made a surprising prediction. Since a changing magnetic field produces an electric field and a changing electric field produces a magnetic field, it followed from the mathematics that there should be electromagnetic waves consisting of oscillating electric and magnetic fields. When Maxwell worked out the speed of the wave, it was about 300,000 kilometers per second. This incredible speed had a deeper significance for a physicist: recent measurements had shown that this was also the speed of light (see Box 2.1). As Maxwell wrote, "We can scarcely avoid the inference that light consists in the traverse undulations of the same medium which is the cause of electric and magnetic phenomena."[22]

Box 2.1

The Imponderable Fluid

One of the first attempts to measure the speed of that imponderable fluid known as light was performed by Galileo. He and an assistant each carried a lantern to the tops of hills which were about a kilometer and a half apart. The idea was that Galileo would flash his lantern, and when his assistant saw the light he would flash his lantern back. Any delay would be due either to the assistant's reaction time, which could be corrected for by first measuring it over a short distance, or to the finite speed of light, which could be calculated by dividing the time by the distance traveled.

Galileo was unable to detect any delay due to the transit of light, so concluded that it was quick. However, unlike most of his contemporaries, he still believed that light traveled at a finite speed rather than transmitting itself instantly from one point to another.

In the mid-nineteenth century, the French scientists Hippolyte Fizeau and Léon Foucault tried more sophisticated versions of the same experiment in which they bounced light pulses off rotating mirrors and measured their deflection. The most accurate estimate by Foucault was 299,796 kilometers per second. The reason Galileo hadn't been able to measure it was because the light's transit time between his two lanterns was only about 0.000005 seconds. When Maxwell worked out his equations and got the same answer for the speed of electromagnetic radiation, he realized it was no coincidence.

In fact, visible light was only one form of electromagnetic wave. As Heinrich Hertz demonstrated in 1888, electromagnetic waves could be produced at will by making an electric current oscillate, for example in a metal antenna. The oscillating current would produce an oscillating electric field (because the electric charge was changing) and also an oscillating magnetic field (because the current was changing). The result was an electromagnetic wave that would speed off into the distance. Hertz believed that his experiments were of no practical use, but they were the basis for the development of radar, radios, TV, and other devices that take their signals from the electromagnetic spectrum.

Of course, one question remained: if there was a wave, where was the ocean? While scientists no longer based their philosophy on Aristotle's earth, water, air, and fire, the fifth element—ether—was still in use. Most Victorian scientists believed that the waves propagated in an invisible medium of ether, whose movements they visualized in terms of complicated assemblies of rods, balls, and hinges. Maxwell derived his theory by thinking about ether in these mechanical terms, even though it doesn't explicitly appear in his equations.

The mysterious, undetectable nature of ether has proved very ponderable over the ages, and it inspired Victorian occultists, artists, and writers such as Charles Howard Hinton, who imagined the solar system as a kind of giant phonograph, scratching out its history in the medium of ether.[23] And although the traditional picture of the ether is today considered a fiction, the idea of space as a medium has never quite gone away. As physicist Robert Laughlin notes, "The modern concept of the vacuum of space, confirmed every day by experiment, is a relativistic ether. But we do not call it this because it is taboo."[24]

Ethereal Vibrations

By the 1890s, scientists believed they had a fairly complete understanding of electricity, and physicists and engineers like Nikola Tesla and Thomas Edison were busy wiring up the United States and other countries with newfangled devices like the light bulb. Scientists were also amusing themselves—and the occasional paying audiences—with an early predecessor of the fluorescent bulb, known as the cathode ray tube. This consisted of an evacuated tube with an electrode at each end. When an electrical charge was applied at the cathode (the negatively charged electrode), it would jump to the positively-charged anode, and a bright, spooky-looking glow would appear where the rays hit the glass around the anode at the far end.

Investigations from a number of scientists revealed that the mysterious rays followed straight lines; could pass through a thin metal foil; cast shadows if a thicker plate was placed in their path; could make a small paddle wheel rotate, as if blowing in the wind; were deflected by magnetic fields; and were apparently unaffected by gravity or, according to some experiments by Hertz, by electric fields. But what were they? Opinions diverged—and as science historian Arthur I. Miller notes, the battle was "fought purely over aesthetics: the aesthetics of waves versus the aesthetics of particles."[25] Scientists on the Continent, and particularly in Germany, believed the rays were some form of "ethereal vibration" (i.e., electromagnetic wave), while British scientists tended to see them as particles. Some, such as the English scientist and vacuum tube designer William Crookes, who like many Victorians was a convinced spiritualist, associated the eerie green glow with the paranormal. The question was finally settled—or so it appeared—in 1897 by J.J. Thomson at the Cavendish laboratory in Cambridge.[26]

Thomson's experimental apparatus was a specially designed twenty-seven-centimeter-long tube, equipped with plates that delivered a strong electric field as well as with a high-quality vacuum. Thomson found that, contrary to Hertz's earlier results, the electric field did deflect the beam towards the positively charged plate, implying that the beam carried a negative charge.[27] It therefore wasn't an electromagnetic wave, but it could consist of what Thomson called "carriers of negative electricity."

Thomson then tried applying a magnetic field to the beam. When charged particles pass through a magnetic field, they are bent at right angles in the direction of the field. Thomson adjusted his apparatus so that the deflections caused by the magnetic and electric fields canceled each other out. From the electromagnetic equations, he could then work out the velocity of the beam. He estimated that it was traveling at about a third of the speed of light.

Once the beam's speed was known, he could then calculate the ratio of charge to mass by measuring how far the beam was deflected by the electric field alone. He attained a charge/mass ratio that was surprisingly high. For comparison, experiments with hydrogen ions (what we now call protons) in electrolytic fluids gave a charge/mass ratio that was about 2,000 times smaller. The most likely explanation—later confirmed by further experiments—was that the cathode ray particles were extremely light.

The results were unaffected by the material of the electrodes—such as aluminum, iron, or platinum—or the kind of gas in the tube, so Thomson concluded that the particle, whatever it was, appeared to "form a part of all kinds of matter under the most diverse conditions; it seems natural therefore to regard it as one of the bricks of which atoms are built up."[28] It was something even more elementary than the elements.

The hypothesis that the electrons, as they were later called, were smaller than the smallest atom, hydrogen, would explain why the beam could pass through a thin metal foil—but it was also "somewhat startling" since it implied that atoms could be divided into even smaller components.[29] As he wrote in 1899, "Electrification essentially involves the splitting up of the atom, a part of the mass of the atom getting free and becoming detached from the original atom."[30] It seemed that atoms weren't so indivisible after all.

Atomic Aesthetics

The cathode ray tube was an early version of a particle accelerator (as well as a precursor to the cathode ray television set). Like microscopes, these accelerators would allow scientists to probe the atom to ever smaller scales—and reveal ever finer levels of detail. The electron was just the first of many subatomic particles to be detected. Democritus's idea of the *atomos*—the smallest unit of matter—has had to be redefined several times.

As discussed in the next chapter, the list of particles soon expanded to include protons and neutrons, which make up the atomic core and most of the atomic weight. Protons carry a positive charge which balances the negative charge of the electron, while neutrons are uncharged. The valency of an atom is determined by the configuration of its electrons, whose harmonies are responsible for the chemical properties of the different elements.

Experiments by Ernest Rutherford and others in the early twentieth century showed that the constituent particles were so small that the atom was mostly empty space. Early models of the atom resembled Copernicus's model of the solar system, with the negatively charged electrons circulating in fixed elliptical orbits around the positively charged core. In place of Newton's inverse-square law of gravity, the system was held together by Coulomb's inverse-square law of electrical attraction.

With accelerators growing in size and power throughout the 1960s and 1970s, exotic new species appeared in the aftermath of collisions. Neutrons and protons were found to be made up of quarks, which come in several different types. Forces were also shown to be mediated by force carrier particles such as the photon, which transmits electromagnetic force.

Today, theorists continue to propose new particles. If supersymmetry is to be believed, then every known particle has a supersymmetric partner. It is of course possible that some of the particles we believe to be fundamental will turn out to consist of yet smaller components.

The search for the building blocks of matter—the integers of nature—was inspired by a desire for order, and by a fascination with patterns like the octave-based structure of chemical valency. Numerical harmony served as both guide and inspiration in the quest to understand the universe. But just as Lavoisier's decomposition of the Aristotelian elements made the world a lot more complicated, so our quest for an elegant theory of matter sometimes seems to be taking us in the other direction—away from simplicity and towards complication.

While atomism may no longer play the calming aesthetic role that it once did in ancient Greece, it certainly helped to *define* an aesthetic which scientists continue to follow. For the Greeks, atoms satisfied the need for something solid, hard, and completely self-contained from which a mechanistic model of physical reality could be built, free of subjectivity or mysticism. That desire is just as alive today and has shaped all branches of science.

Ironically for an approach which described secondary, subjective qualities as unreal, atomism would come to characterize the look and feel of what is considered "hard" science. Consider, for example, the following antitheses, taken from a list compiled in 1985 by the philosopher Mary Midgley:

hard	soft
determinism	free will
mechanism	teleology
skepticism	credulity
reason	feeling
objective	subjective
realism	reverence
specialism	holism
prose	poetry
male	female
clarity	mystery

As Midgley pointed out, these concepts provided scientists with what she calls a "mental map . . . marked only with the general direction 'keep to the left'."[31] Atomism is the archetype of a hard, deterministic, mechanistic, objective, scientific theory (we will consider the male/female aspect in Chapter 5). It is the heart of the "mechanistic philosophy," defined by quantum physicist David Bohm as the belief that "the great diversity of things that appear in all of our experience, every day as well as scientific, can all be reduced completely and perfectly to nothing more than consequences of the operation of an absolute and final set of purely quantitative laws determining the behaviour of a few kinds of basic entities or variables."[32] Accelerator projects such as the Large Hadron Collider are just the most graphic examples of this approach, which owes its heritage to Democritus and the search for integrity.

This mechanistic philosophy is tied up with the desire for precision and logical exactness. As author Steve Talbott puts it, "We want to be able to say, 'I have exactly this—not that and not the other thing, but this.' The ideal of truth at work here is a yes-or-no ideal. No ambiguity, no fuzziness, no uncertainty, no essential penetration of one thing by another, but rather precisely defined interactions between separate and precisely defined things. We want things we can isolate, immobilize, nail down and hold onto. . . . Whatever looks complex and of diverse nature must be analyzed into distinct, simple parts with clearly spelled-out relations."[33] In the same way that elements must have integrity, so any statement we make about the world should be true or false, not a mix of the two. As discussed further in Chapter 10, this insistence on integrity comes at a considerable cost.

Atomism, in the sense of a tendency to break the world down into separate and independent elementary components, has affected far more than science. In architecture, our living arrangements have become ever more atomized (see Box 2.2). The modern idea that art objects are autonomous, self-contained, and independent of their moral or social context would have been incomprehensible to someone in the Middle Ages.[34] In education and work, expertise is parsed into ever finer categories. But we are beginning to realize that at some point atomism loses its utility. Lavoisier showed that air is mostly nitrogen and oxygen, and that water is hydrogen and oxygen, but this tells us little about the dynamics of a cloud. We know that salt is made from sodium metal and chlorine gas, but that doesn't help a cook judge its effect on flavor. As discussed later, attributes as basic as the solidity or electrical conductivity of an object are collective emergent effects that can only be understood for a large number of atoms, and are often robust to or even reliant on chemical impurities. At the social level, everyone from advertisers to social revolutionaries is becoming increasingly attuned to the importance of networks. Integrity is a property, not of individual parts but of their organization into a coherent structure.

Scientists such as Lavoisier, Dalton, Mendeleev, and Thomson may not have been consciously guided by aesthetic considerations, other than a keen appreciation for numerical patterns. In many ways they were just following a historical trajectory that was largely driven by technology. But today, the continued emphasis on breaking things into pieces does appear to reflect aesthetic choices. As physicist Anthony Zee observes, "aesthetics has become a driving force in contemporary physics"—and that may be something of an understatement.[35] In the next chapter, we will consider another aesthetic property that has remained remarkably stable over millennia—that of stability.

Box 2.2

The Atomized Society

Can abstruse scientific theories about invisible particles affect the way we live, or vice versa? It is interesting to compare the historical trajectory of atomism in science with the individuation of society as expressed through language and living arrangements.

In the Archaic Age of Greece (750–500 BC), the idea of an individual human self was—as with an infant—still in its infancy. In Homer, a person's soul or personality only made an appearance on death, when it was described as leaving the body. It was not until the seventh century BC that lyric poets such as Archilochus begin to express their feelings in the first person singular.

Typical homes in ancient Greece were based around a courtyard, which gave privacy from the outside but not from the other inhabitants of the house. Philosophers such as Democritus probably did some of their work at night, when the house was quiet. In ancient Rome, most houses were open to the street, while the wealthy lived in courtyard houses and retired to their villas when they needed privacy.

In the early Middle Ages, European homes were similar to barns, with a roof open to the ceiling and a large hearth. Richer households added extra rooms to the periphery of this large hall, but it continued to function as a semi-public space with frequent guests, visitors, random passersby, various pets such as hunting birds or dogs, and so on. Meals were a rowdy affair and it was considered rude to seek privacy.

In sixteenth- and seventeenth-century England, multi-room cottages became the norm even for laborers. A typical cottage had a general living area, a shared room for sleeping, and a storage area called the buttery. Those higher up the social ladder, such as Isaac Newton's yeoman father, had several more rooms including servant's quarters and separate bedrooms. At the same time that Newton was describing matter in terms of hard, independent particles, there was, as philosopher and geographer Yi-Fu Tuan observes, a burgeoning social desire for "more specialized space, for a room or two in which a measure of privacy was possible."[36]

Homes continued to become ever more private and contemplative places. Wall mirrors became common in the late seventeenth century, and libraries were a popular addition in the early eighteenth century. Words such as "ego," "character," "self-love," and "melancholy" began to be used in their modern sense, and the one-letter word "I" appeared with increasing frequency. It is perhaps unsurprising that the late-nineteenth-century triumph of atomism coincided with the development of psychoanalysis.

In 1937 the author Ayn Rand, whose individualistic philosophy was hugely influential, wrote in her novel *Anthem* (her working title was *Ego*), "And now I see the face of god . . . This god, this one word: 'I.'"[37] Her follower and lover Nathaniel Branden—who unselfishly changed his last name, Blumenthal, to include hers—went on to write some twenty books on psychology, ten of which had "self" in the title. The *principium individuationis* (principle of individuation), for which Nietzsche called Apollo the archetype, was in full flower.[38]

This atomizing trend has continued—a 1995 survey showed that most Americans saw others (but not themselves) as "increasingly atomized [and] selfish."[39] As Tuan notes, "Western culture encourages an intense awareness of self and, compared with other cultures, an exaggerated belief in the power and value of the individual. . . . This isolated, critical and self-conscious individual is a cultural artifact. We may well wonder at its history. Children, we know, do not feel and think thus, nor do nonliterate and tradition-bound peoples, nor did Europeans in earlier times."[40]

Of course there are many factors involved, but given the importance of social and scientific narratives, and the tendency to align physical spaces with mental spaces, perhaps our social atomism has something to do with our belief in atoms—and the recent Internet-driven resurgence of social networks may be related to the shift in science from reductionism to complexity.

3

Radiance

Geometry is unique and eternal, a reflection of the mind of God. That men are able to participate in it is one of the reasons why man is an image of God.

Johannes Kepler, quoted in Arthur Koestler, *The Sleepwalkers*

The elementary particles have the form Plato ascribed to them because it is the mathematically most beautiful form. Therefore the ultimate root of phenomena is not matter but instead mathematical law, symmetry, mathematical form.

Werner Heisenberg, *Die Plancksche Entdeckung und die philosophischen Grundfragen der Atomlehre*

Equilibrium, in its very best sense—in the sense the Greeks *originally* meant it—stands for the strange spark that flies between two creatures, two things that are equilibrated, or in living relationship. It is a goal: to come to that state when the spark will fly.

D.H. Lawrence, "Him with His Tail in His Mouth"

The desire for stability is an eternal human trait, and nowhere more than in science. Plato saw the world as an imperfect reflection of abstract forms that exist outside of time. Only mathematical constructions such as lines and circles were "eternally and absolutely beautiful." Today, perhaps the closest thing to Platonists are the scientists who search for the eternal, unchanging mathematical laws which they believe govern the universe. In contrast, nature itself is in a constant state of flux—in fact, it is positively explosive.

This chapter charts our understandings of movement, energy, and radiation, and explores the tension between the forces that drive change in the universe and our desire for a stable and consistent model.

Democritus's theory of the atom was formulated partly in response to philosophical puzzles, such as those raised by Zeno, about the nature of movement. Zeno's teacher was Parmenides, a priest of Apollo who authored a poem called *On Nature* in which he journeys to the abode of a goddess, sometimes identified with nature. The goddess there instructs him on both the true reality, referred to as What Is, and the opinions of mortals.

According to the goddess, the What Is must be "ungenerated and deathless, whole and uniform, and still and perfect."[1] Since it is "now together entire, single, continuous," it must have been the same for all time; for otherwise, "what birth will you seek of it?" She concludes that "it must either be altogether or not at all . . . thus generation is extinguished and destruction unheard of."

The What Is must also be completely stable and unchanging, "remaining the same, in the same place, and on its own, it rests, and thus steadfast right there it remains; for powerful Necessity holds it in the bonds of a limit, which encloses it all around." It is completely uniform and symmetric, "perfected from every side, like the bulk of a well-rounded globe, from the middle equal every way. . . . Since it is all inviolate, for it is equal to itself from every side, it extends uniformly in limits."

The What Is can only be apprehended through logos, or reason, since sensory perceptions are deceptive. The cosmos which we actually perceive therefore has "no genuine conviction."

Only fragments of the poem remain, so its meaning is debated. The traditional interpretation has been to say that Parmenides literally thought that all movement is an illusion. Saying that an object appears in a new location is equivalent to saying that matter must magically be created in this new place, which is impossible because the What Is cannot be created or destroyed.

This obviously seems to be an unhelpful conclusion, and it is interesting that Parmenides devotes much of the poem to a detailed explanation of cosmology, which would appear somewhat pointless if what we see and experience were really no more than an illusion. So an alternative interpretation might be to say that the What Is really refers to some underlying laws or principles of the universe. In this case, his statement that they are stable, eternal, uniform, rational, and symmetric would be a pretty good summary of our modern scientific view.

In his poem, Parmenides traveled to visit his goddess in a chariot, escorted by the maiden daughters of the sun-god Helios. Today, particle physicists just log on to their account at the LHC, but the aim is still the same: to find the What Is.

Good Form

In his dialogue *Parmenides*, Plato gives a fictionalized account of a meeting between Parmenides, his student Zeno, and a young Socrates, in which they thrash out heady philosophical issues such as the Theory of Forms. This theory states that material objects are imperfect versions of abstract Forms, which exist independently of time and space. Included in these Forms are mathematical and aesthetic concepts such as unity, beauty, and goodness, the last of which Plato identifies in *The Republic* as the highest of all Forms.

As with the What Is of Parmenides, the Forms can only be apprehended through reason. When mathematicians deal with a geometrical figure such as a line, they are not thinking of the kind of lines one can actually see or draw, "but of the ideals which they resemble . . . which can only be seen with the eye of the mind."[2] A real line, say one drawn by a carpenter, is but a flawed reproduction of the real thing. True beauty is limited to things such as "straight lines and circles, and the plane or solid figures which are formed out of them by turning-lathes and rulers and measures of angles; for these I affirm to be not only relatively beautiful, like other things, but they are eternally and absolutely beautiful, and they have peculiar pleasures, quite unlike the pleasures of scratching."[3]

Under interrogation by the fictional Parmenides, Socrates is a little unsure on whether more down-to-earth objects such as mud also have associated Forms (one imagines him scratching his beard at that one). This ambiguity was related to a more general question, which was the relationship between abstract, ethereal, pure ideas and messy, pluralistic reality.

Plato left this question unresolved—in *Parmenides*, he seems to be admitting that the Theory of Forms was not entirely flawless and perfect itself—but it was addressed neatly by Aristotle, who simply divided the cosmos into two parts. The sublunary sphere was everything inside the sphere of the Moon, including the Earth. All matter within this sphere was made up of the four elements. Because the elements could move and transmute into one another, this region was subject to mutability and change, but still yearned for equilibrium. Everything outside the Moon's sphere, in contrast, belonged to the perfect, eternal, unchanging domain. Things there were constituted entirely from ether. The only kind of movement permissible was circular: the spheres of the planets rotate, but never change.

In his *Physics*, Aristotle presented a detailed theory of how and why objects move. Each of the four elements strives towards its natural place, with earth at the center, surrounded by water, then air, and fire pressing up towards the heavens. The weight of a material therefore depends on its composition. Something with a high fire content, such as smoke, will be lighter than air and will therefore tend to rise up, while things with a high earth content will tend to sink down. This type of vertical, up-down motion reflected the desire for objects to achieve their natural place in the cosmic order and become more perfect.

Aristotle argued that the speed of motion depends on an object's weight—with heavy ones falling faster than light ones—and the density of the medium, so for example a ball will fall faster in air than it will in water. In a vacuum, he believed, objects would fall infinitely fast, which was another reason for it not to exist.

Horizontal motion was different because it did not reflect an object's tendency to return to its natural state and was therefore, in a sense, unnatural. Objects only moved this way if they were pushed. This was easy to understand for things which were being constantly propelled, such as a person walking or a wagon being pulled by a horse, but was more problematic for something like a thrown projectile, which kept going even after it was released. Aristotle proposed that the movement was somehow maintained by a process named antiperistasis: as the object moves forward, it displaces air, which rushes in behind it and propels it along. Alternatively, the throwing process might set the air in front of the projectile moving in such a way that it drags the projectile along behind it.

Aristotle's theory of motion, therefore, was actually a theory of stability. Everything had its natural place in the cosmos and did not move from it unless forced. In the sublunary sphere, objects were subject to motion only because the four elements were mixed up in them and could transmute, so stability and perfection could not be attained. In the heavens, the celestial bodies were trapped in their crystalline spheres of ether, endlessly repeating the same choreographed dance.

Aristotle's emphasis on stability was partially motivated by the fact that most objects did seem to come to a halt when the force propelling them was removed (it has been said that if ancient Greece had been blessed with smoother roads and better wheels, he would have come to a different conclusion). But it also reflected a broader aesthetic view of the cosmos as an inherently ordered, stable, and beautiful place. Change and mutation were seen as signs of imperfection. After all, if something were perfect, why would it want to change?

This dislike of change also extended to Greek society, which was structured in a similarly static, hierarchical manner as the model of the cosmos. At its base were slaves, who Aristotle saw as "totally devoid of any faculty of reasoning" but who did most of the work.[4] Then, in ascending shells, were ex-slaves, foreigners, and artisans, and finally the land-owning, non-working upper class. They alone could be citizens, and they oversaw everything from above, like the stars in the firmament. Women did not take part in political life and took their social class from their male partner.

Karl Marx said that "the ruling ideas of every age are always the ideas of the ruling class."[5] A model of the universe which suggested that each object, and each class, has its natural place in the cosmic scheme would certainly have appealed to the male leisure class that ruled ancient Athens.

Class Structure

Much later, the static philosophy proved attractive to the medieval church for the same reasons, and they adopted it with gusto. In the early fifth century, Macrobius summarized Christian cosmology as follows: "Since, from the Supreme God Mind arises, and from Mind, Soul, and since this in turn creates all subsequent things and fills them all with life, and since this single radiance illumines all and is reflected in each, as a single face might be reflected in many mirrors placed in a series; and since all things follow in continuous succession, degenerating in sequence to the very bottom of the series, the attentive observer will discover a connection of parts, from the Supreme God down to the last dregs of things, mutually linked together and without a break. And this is Homer's golden chain, which God, he says, bade hang down from heaven to earth."[6]

This view of a cosmos united by a divine radiance was essentially a reengineered version of the Greeks' static, hierarchical world view, with Plato's Form of the Good transposed to the Christian God. Thomas Aquinas later even used Aristotelian physics to prove the existence of God. If objects only move because another object pushes them, like balls on a billiard table bumping into one another, then there must have been a "prime mover" who took the first brilliant shot that got all the balls in motion.

While Aristotelian physics was accepted as dogma by the Church, it did not completely escape criticism. The process of antiperistasis, in particular, seemed a little contrived, even to the philosopher himself. In 500 AD, John Philoponus wrote a *Commentary on Aristotle's Physics* in which he argued that an "incorporeal motive force" must be imparted to a projectile in order to keep it moving.[7] However, such technical problems were generally suppressed or ignored—at least until 1572, when it became very obvious that the heavens were less immutable than previously thought.

On the evening of 11 November that year, the astronomer Tycho Brahe was leaving his alchemical laboratory when he noticed that the constellation of Cassiopeia had a new addition. The new star was as bright as Venus and caused a huge stir across Europe—especially when Brahe later proved, using detailed observations, that it was definitely located in the "firmament" as opposed to the sublunary sphere. He wrote up his results in his 1573 book *De Stella Nova*.

Such supernovas are actually massive stellar explosions, which release a burst of radiation lasting months—or, in this case, years. Similar events are known to have occurred several times in recorded history, and it is therefore strange that European astronomers had not taken note of them earlier. Supernovas in the years 269, 286, 293, 1006, 1054, and 1181 would have been visible to anyone in the northern hemisphere, and were observed in the East by Asian astronomers. So why was it that Brahe could see the supernova in 1572, and Kepler could observe another in 1604, but until then they had been invisible? As the mathematician Ralph Abraham wrote, "We have to conclude that the astronomers of medieval Europe were effectively blinded by their faith in Aristotelian dogma."[8] It is an example of how models can close our minds to new information by creating cognitive dissonance. On the other hand, Copernicus's Earth-centered model may have given astronomers permission to see the unthinkable.

In 1621, Kepler revised his *Mysterium Cosmographicum* and replaced the word soul (*anima* in Latin) with force (*vis*) to describe the "motive cause" which propels the planets around the Sun. For perhaps the first time, scientists were thinking that the cosmos were ruled by dynamical principles rather than a purely static order.

Kepler's ideas about cosmic harmony were reinforced when he read a book by Vincenzo Galilei on musical harmony.[9] It was Vincenzo's son, Galileo Galilei, who would take the first steps towards building a new theory of motion based on dynamics.

Accelerator

Aristotle had stated that heavy objects fall faster than light objects. While this law was certainly true for particular objects—a feather falls more slowly than a hammer—the reason, we now know, is because light things are more affected by air resistance.

The Apollo 15 astronaut David Scott demonstrated this fact by dropping a feather and a hammer on the airless surface of the moon, where they both fell at the same rate. According to legend, Galileo tested the law—less spectacularly, but at less expense to the taxpayer—by tossing stones and other objects of different weights off the Leaning Tower of Pisa. But he also performed a much more detailed experiment, using what might be described as the world's very first accelerator.

The "accelerator" was an inclined plane, about seven meters long and thirty centimeters wide; the measuring device was a water clock; and the object to be accelerated was a "hard, smooth, and very round bronze ball."[10] As Galileo wrote in his 1638 work *Dialogues Concerning Two New Sciences*, the plane contained a groove "a little more than one finger in breadth" for the ball to roll down.[11] The groove was lined with parchment, and was as "smooth and polished as possible."[12] For timing, Galileo employed "a large vessel of water placed in an elevated position; to the bottom of this vessel was soldered a pipe of small diameter giving a thin jet of water, which we collected in a small glass during the time of each descent . . . the water thus collected was weighed, after each descent, on a very accurate balance; the difference and ratios of these weights gave us the differences and ratios of the times."[13]

The purpose of the inclined plane was to slow the ball's falling motion down to the point where it could be accurately measured using the primitive water clock. Using this simple apparatus, Galileo discovered that a falling body experienced uniform acceleration: if it began from rest, then after two seconds it would be moving twice as fast as after one second. The distance traveled varied with the square of time, so after two seconds it would have traveled four times farther than it had at one second.

Again, the speed did not depend on the weight of the ball. This implied a somewhat subtle but deep symmetry which Einstein later called the equivalence principle: the inertial mass, which resists acceleration, is the same quantity as the gravitational mass, which experiences the force.

Since balls rolling down a ramp experienced uniform acceleration, Galileo wondered if a similar thing would happen if a ball rolled the other way, up the ramp. Sure enough, he found that the ball experienced a constant deceleration (or negative acceleration) as it rose. Galileo reasoned that, if the acceleration is positive when the ramp is pointed down and negative when it is pointed up, then when the ramp is horizontal there should be zero acceleration along it. In other words, a ball will keep moving at the same speed unless impeded by something else, such as friction. Aristotle had taught that objects seek a stable position; Galileo showed that it is actually the object's velocity which "wants" to remain stable.

Galileo's work, with its use of carefully controlled experiments to deduce fundamental physical laws, in many ways marked the beginning of modern science. Thomson's experiments at Cavendish a quarter of a millennium later would again involve the careful measuring and balancing of forces, though there the bronze balls were replaced by electrons.

Conservation Laws

Galileo was threatened with torture and put under house arrest for his un-Aristotelian views, in particular his agreement with Copernicus that the Earth is not the center of the universe. But less than forty-five years after his death, Isaac Newton published his *Principia*, which showed that Galileo's rolling balls and Kepler's ellipses were two manifestations of the same phenomenon measured over vastly different scales.

In addition to the law of gravity, Newton's *Principia* proposed three laws of motion. The first law, also known as the law of inertia, stated that, "Every body perseveres in its state of rest, or of uniform motion in a right line, unless it is compelled to change that state by forces impressed thereon."[14] The second read, "The alteration of motion is ever proportional to the motive force impressed; and is made in the direction of the right line in which that force is impressed." These two laws were essentially a generalization of Galileo's findings with his acceleration experiments, in which the motive force of gravity was seen to cause a uniform acceleration.[15] The third law said, "To every action there is always opposed an equal and opposite reaction." Forces therefore represented a symmetrical interaction between separate bodies. When a horse pulls on a cart, the cart pulls back. When the Earth pulls a metal ball down a ramp, the Earth feels a force as well, but its huge mass means there is little effect.

The *Principia* also contained two conservation principles. One was the conservation of mass. In a closed system in which nothing is allowed to enter or escape, Newton assumed that, while one substance might conceivably transmute into another, the total amount of mass should remain constant. The mass conservation principle was explicitly stated, and experimentally demonstrated, by Lavoisier (see Chapter 2). The second conservation principle, which Newton showed was a consequence of the second and third laws of motion, was that when two or more bodies interact, their total momentum remains constant.

The momentum of an object is a so-called vector quantity, with magnitude equal to the object's speed multiplied by its mass, oriented in its direction of travel. If two identical objects—say protons in the LHC—are moving at the same speed but in opposite directions, then their momenta are of different signs, so they sum to zero. According to the conservation of momentum, after the collision of these objects the total momentum of whatever comes out must still be zero.

A similar principle had been proposed by the French philosopher René Descartes, who in his 1644 *Principia philosophiae* wrote that "God is the primary cause of motion; and he always preserves the same quantity of motion in the universe. . . . For we understand that God's perfection involves not only his being immutable in himself, but also his operating in a manner that is always utterly constant and immutable."[16] However, Descartes identified his "motion" as the product of mass and speed, which corresponds to the magnitude but not the direction of momentum.

In the same way that Newton's law of gravity replaced the Greek circle model with a model based on another kind of circle—the symmetric law of gravity—so his laws of motion introduced a different kind of symmetry and stability, which applied to the underlying equations rather than the observed phenomenon. Instead of an object's position being conserved, it was abstract quantities like mass and momentum. The Platonic Forms were being resurrected, this time in the form of mathematical equations.

Saving Energy

Conservation laws continued to play an important role in the development of physics. Perhaps the most important was the law of energy conservation, which took some two hundred years to develop. Newton's contemporary, Gottfried Wilhelm Leibniz, had noticed that a quantity he called the *vis viva*, or living force, tended to be conserved in certain mechanical systems. Later renamed kinetic energy, this force corresponds to the energy gained by the ball rolling down its ramp in Galileo's experiment.

The source of the ball's kinetic energy is gravitational force. This knowledge led to the concept of potential energy. At the top of the ramp, the ball starts with an endowment of gravitational potential energy, which is converted to kinetic energy as it descends. To lift the ball back up requires another kind of energy we are all familiar with, namely work. Other types of forces, such as magnetic or electrical attraction, also produce potential energy.

Another type of energy was the thermal energy stored in heat. According to the kinetic theory of heat, which was developed in the early nineteenth century, this was associated with the average kinetic energy of atoms as they randomly jostle around. At normal room temperature, air molecules move at an average speed of more than 1,500 kilometers per hour. When warmed, atoms move more violently and push against one another as they collide (which is why substances such as the mercury in a thermometer tend to expand as they warm up).

Despite these discoveries, energy still remained a somewhat nebulous concept, and it wasn't clear how its different forms were connected. In 1845, the English physicist and brewer James Prescott Joule, who had been a student of the atomist John Dalton, managed to quantify the relationship between gravitational potential energy and heat energy.[17] His experiment involved a weight which was set up to turn a paddle wheel when it fell. The paddle wheel was immersed in an insulated barrel of water. By measuring the temperature of the water, Joule could thus measure exactly how much heat was generated by the falling weight. The idea that mechanics, heat, electricity, magnetism, and light were all different aspects of a single type of energy—an energy that was at all times conserved—was stated in 1847 by the German physician and physicist Hermann von Helmholtz.[18]

In 1915, the German mathematician Emmy Noether made the remarkable discovery that conservation principles were related to statements about symmetry.[19] Conservation of momentum is a consequence, it turns out, of the statement that a physical experiment will produce identical results independent of spatial position—in other words, the laws of physics will not change if you take a step to the side. Conservation of energy is a consequence of saying that experimental results do not depend on the time: we can repeat Galileo's experiments today, or the day after tomorrow, and get the same result. As discussed further in Chapter 7, symmetry and stability are like two sides of the same coin.

In fact, physical laws of any type, as physicist John D. Barrow points out, "are equivalent to invariance principles, that is, statements to the effect that some entity does not change."[20] This raises a question: if the world is run by stability laws, what causes transformation and mutability? If physics is based on conservation, then what creates energy and warmth in the first place? And if stability is beautiful, does that mean change is ugly?

Radiance

The question of warmth creation was much on the mind of Marya Sklodowska after she moved in 1891 from her native Poland to study at the Sorbonne.[21] As she later wrote, "The room I lived in was in a garret, very cold in winter, for it was insufficiently heated by a small stove which often lacked coal. During a particularly rigorous winter, it was not unusual for the water to freeze in the basin in the night; to be able to sleep I was obliged to pile all my clothes on the bedcovers."[22]

Fortunately she was warmed by what D.H. Lawrence called "the strange spark that flies between two creatures" and in 1895 she married Pierre Curie. Two years later, Marie Curie—as she was now known—became one of the first women in Europe to embark on a PhD. Perhaps out of an understandable desire to find an infinite source of heat, she chose as her research topic the mysterious phenomenon of radioactivity (a name she coined) by which materials like uranium seemed to produce energy from nothing.

The French physicist Henri Becquerel had recently discovered that uranium salts, which glow in the dark after being exposed to sunlight, also produce invisible rays which can expose photographic plates even when they have been left in the dark and aren't glowing. The most puzzling aspect of the phenomenon was that it seemed to violate the principle of energy conservation. Where was the energy coming from to produce these rays, if it wasn't the Sun?

Curie soon discovered that the uranium ore known as pitchblende was actually more radioactive than uranium itself. She concluded that it must contain some other highly radioactive substance. Working together with husband Pierre, she eventually succeeded in isolating not one but two new sources of radiation. One they called radium and the other—in a political gesture to her homeland, which was still under Russian rule—they called polonium. Both radium and polonium were present in only trace quantities—a fraction of a gram in tons of pitchblende—but they were hundreds of times more powerful than uranium. Pierre calculated, for example, that a lump of radium could heat more than its weight in water from freezing to boiling in one hour—not just once, but over and over.

"One of our joys," Marie wrote later, "was to go into our workroom at night. We then perceived on all sides the feebly luminous silhouettes of the bottles or capsules containing our products. It was a lovely sight and always new to us. The glowing tubes looked like faint fairy lights."[23] Their discoveries brought the Curies fame, but also disaster. In April 1906, Pierre was killed by a horse-drawn wagon after he slipped and fell in the street. He had been suffering from dizzy spells that were likely caused by radiation poisoning. Marie died in 1934 from leukemia, which may also have been the result of working with radioactive substances without proper protection. Her laboratory notebooks are still radioactive and are kept in a lead-lined vault. (The dangers of polonium were demonstrated more recently when it was used in the 2006 poisoning of the former Russian spy Alexander Litvinenko. A tiny quantity put in his food or drink was enough to slowly destroy his internal organs.)

We often think of radiation as being artificially produced by scientists. However, our bodies are constantly bombarded with naturally-occurring radiation from both the earth and the sky. The ground contains radioactive substances like uranium and thorium, and the heavens contain the radiation of exploding stars. Curie's discovery of radium made the topic of radiation impossible for the scientific community to ignore, and would lead to a reassessment of the laws of physics—and the idea of stability.

Transmutation

Another scientist intrigued by the properties of radiation, but who apparently managed to avoid its harmful side-effects, was the New Zealand physicist Ernest Rutherford. A student at Cambridge under J.J. Thomson, he moved in 1898 to McGill University in Montreal. Working with the English chemist Frederick Soddy, Rutherford discovered that radioactivity involved nothing less than transmutation: the mysterious process which had been proposed by Aristotle but had long eluded alchemists. Radium, they found, emits radioactive particles, which Rutherford dubbed alpha particles. In doing so, the radium transmutes into radon gas. Every 1,602 years, half the radium gets transformed in this way. After another similar period, radium's so-called half-life, half the remaining radium is lost, and so on. After a million years only a trace will remain. The energy is therefore associated with the transformation of matter from one form to another. This seemed to go up against one of the key aesthetic principles of science—namely the idea of atoms as eternal, immutable objects.

"For Mike's sake, Soddy, don't call it transmutation," said Rutherford. "They'll have our heads off as alchemists."[24] Indeed, Curie disagreed with Rutherford and Soddy because she believed that "atoms were by their very essence unchangeable."[25] One of the reasons why atomism had never been supported by the Church was because it was incompatible with the doctrine of transubstantiation, which said that Communion bread and wine transform into the flesh and blood of Christ.[26] But Rutherford believed that it was not atoms which were immutable—it was their constituents, together with the laws of physics. Atoms could break apart and reconfigure themselves, but these reactions were still subject to principles such as energy conservation.

In 1907, Rutherford returned to the UK to take up a position in Manchester, where he set up a group to research the topic of radiation. Using similar methods as those developed by J.J. Thomson to analyze the electron, supplemented by new detector methods (see Box 3.1), he found that alpha particles were the same as helium atoms stripped of their two electrons. His main interest, however, was using the alpha particles as a tool to probe the atom's structure.

The prevailing model of the atom was J.J. Thomson's "plum-pudding" model, which assumed that charge was more or less uniformly distributed in space. While at McGill, Rutherford had found that alpha particles were resistant to deflection by magnetic or electrical forces, but could be scattered by a sheet of mica crystal less than a thousandth of a millimeter thick. It seemed that the forces in the mica were far stronger than the forces that could be imposed externally using magnets or electric fields. In 1909, he asked Geiger and a student called Ernest Marsden to repeat the scattering experiment, this time using a thin sheet of gold foil instead of mica and a scintillation screen as detector. And, rather than just placing the scintillation screen behind the foil, they tried placing it to the side to see if the forces were so great that they could create really large deflections.

Counting the scintillations was hard work. Marsden and Geiger had to spend many hours in a darkened room, taking turns every minute or so to peer at the scintillation screen through a microscope and record the flashes. From hundreds of thousands of observations, they were amazed to find that, while most of the alpha particles passed right through the gold foil, about one in 8,000 would bounce straight back off it. As Rutherford later put it, "It was as if you fired a fifteen-inch artillery shell at a piece of tissue paper and it came back and hit you."[27]

The implication seemed to be that matter was mostly empty space, with minute concentrations of positively charged material that repelled the alpha particles. This was an astonishing idea: even solid gold, it seemed, had no real substance, but was only an illusion created by a web of interacting forces. If matter was made up of atoms and the void, as Democritus had said, then it was mostly void. Again, this posed an aesthetic challenge; it suggested that what counted was less the constituents of the atom than the forces and relationships between them. Properties such as stability and solidity emerged, it seemed, from the equations which governed matter, rather than matter itself. The artist Wassily Kandinsky later wrote how this insight helped him to overcome an artistic block: "The collapse of the atom was equivalent, in my soul, to the collapse of the whole world. Suddenly the thickest walls fell. I would not have been amazed if a stone appeared before my eye in the air, melted, and became invisible."[28]

Box 3.1

Detectors

In September 1894, the Scottish physicist Charles Wilson was working at a meteorological observatory on Ben Nevis, the tallest of the Scottish peaks. He was intrigued by atmospheric effects caused by the diffraction of light in cloud droplets, such as coronas and glories. A corona is a ring of colored rays around the Sun, while a glory is a ghostlike halo effect—known more evocatively as the Brocken Specter—that on rare occasions appears around your shadow when you look away from the Sun. Wilson was fascinated by such phenomena, and it occurred to him that it should be possible to reproduce them in the laboratory.

The next year, at the Cavendish laboratory, he succeeded in making his own clouds in a jar. The "cloud chamber" consisted of a glass chamber full of water vapor, with a piston at one end. When the piston was rapidly withdrawn, the gas inside the chamber expanded and cooled. The water vapor then condensed on tiny dust particles, forming a mist. With the right settings, the droplets could reproduce some of the effects Wilson had seen on Ben Nevis. He also found that droplets would form on charged molecules, known as ions, in the air. Rutherford's team used the cloud chamber to photograph the gracefully spiraling vapor trails left by ions, such as alpha particles, as they deflected under electric or magnetic fields.

As discussed later, cosmic rays from outer space were also capable of leaving traces in cloud chambers. Scientists have recently discovered that cosmic rays cause the formation of real clouds in the atmosphere, and can affect the climate.[29] The glimpses into the world of subatomic physics afforded by the cloud chamber, as well as its descendant the bubble chamber (see Chapter 6), provide some of the most hauntingly beautiful images in physics. They are the scientific equivalent of street photography, where patience, technique, and an eye for the serendipitous random encounter are all required.

The scintillation screen was invented by William Crookes, who also invented the Crookes tube mentioned in Chapter 2. One day in 1903, Crookes had mislaid a small speck of radium that he was working with. He knew that radiation caused zinc sulphide to fluoresce, so he took a piece and moved it around his work area until it began to glow. He then decided to take a closer look at the fluorescence through a magnifying glass. Instead of a steady glow, he saw that the substance was giving off a marvelous display of individual flashes that seemed to shoot out at random. He deduced that the separate sparkles were produced by individual particles.

Crookes turned his discovery into both a particle detector and a device called a Spinthariscope, from the Greek *spintharis* for spark. This consisted of a tube with a zinc sulphide screen at one end, a viewing lens at the other, and a miniscule speck of radium salt near the screen. The Spinthariscope was for a while a popular amusement among the Victorian upper classes, who weren't worried about staring at radioactive material.

Together with the Geiger counter, invented by Rutherford's recruit Hans Geiger, these devices were the prototypes for the massive detectors used in modern colliders.

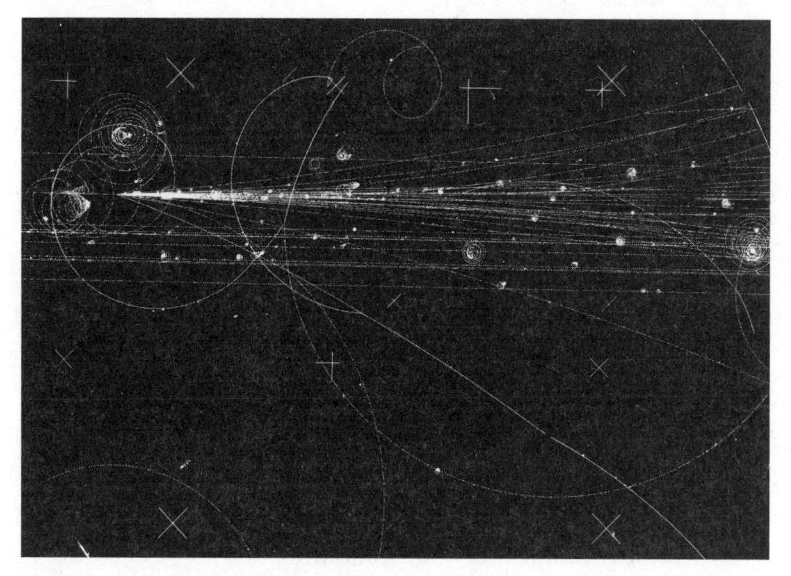

Figure 3.1 Particle tracks in a bubble chamber.[30] Source: High energy pion interaction event (Courtesy of Fermilab Visual Media Services)

Atoms and the Void

It took Rutherford a while to work out the details, but in 1911 he proposed his model for the atom, in which a cloud of negatively charged electrons circulated around a very small, positively charged nucleus like planets around a sun. Alpha particles brushed easily past the extremely light electrons, but if one by chance was headed for the far-heavier nucleus it would be deflected by the charge. Rutherford even simulated the process by swinging an electromagnet, pendulum-fashion, from a nine-meter wire and directing it towards a second electromagnet on a bench, oriented in such a way that it repelled the first.[31] Since he knew the charge and energy of the alpha particles, he could deduce from the scattering mechanics and the conservation of energy that the radius of the nucleus was about 10^{32} centimeters, or one ten-thousandth of the atomic radius.

Rutherford later found that when nitrogen gas was bombarded with alpha particles, the nitrogen atoms gave up a hydrogen atom that had been stripped of its electron, which he called a proton. In another example of transmutation, the loss of a proton in the nitrogen atom turned it into oxygen. Rutherford deduced that it was protons that gave the nucleus its positive charge.

This model of the atom—with the lightweight, negatively charged electrons buzzing around a heavy, positively charged core—was intuitive, easy to understand, and pleasingly concrete: you could even rig up a toy version using magnets and wire. It still corresponds to most people's mental picture of an atom. However, the model had a number of problems.

One was that the atomic weight of many elements could not be explained by the protons alone. For example, hydrogen consisted of an electron and a proton, so it had an atomic weight of one; while the next-heaviest element, helium, had two electrons, so for it to be electrically neutral it needed to have two protons. However, its atomic weight was not two, but four—so something had to be providing the missing mass.

Rutherford therefore hypothesized the existence of a neutral particle in the nucleus, which he called the neutron. He initially believed that the neutron was a doublet consisting of a proton and electron bound together. This would explain how high-energy electrons appeared to be emitted from the nucleus in beta radiation (see Box 3.2). But when he tried to explain the mechanics to an audience in 1927, few were convinced. As a student wrote, "The crowd fairly howled. I think Rutherford came nearer to losing his nerve than he ever did before."[32] The fact that the neutron was not electrically charged made it very hard to detect. The puzzle was eventually solved by Rutherford's student, James Chadwick (see Chapter 5).

Another question was what held the nucleus together. Some sticky, short-range force—appropriately known as the strong force—had to be counteracting the electromagnetic repulsion of the protons. It was later found that the strong force acts directly on the quark components that make up protons and neutrons. The neutrons are uncharged, but contribute a strong attraction which generally helps stability.

An even more challenging problem with the solar system model was that it was dynamically unstable. In the real solar system, the Earth is constantly being accelerated towards the Sun, not in terms of speed, but in the sense that the Sun directs the Earth to follow an elliptical path rather than shooting off in a straight line. In Rutherford's model of the atom, the electrons would experience a similar accelerating force from their attraction to the positive nucleus. But according to Maxwell's equations, an accelerating electron would produce an electromagnetic wave, just as accelerating electrons in an antenna create a radio signal. The electrons would therefore radiate away their energy, slow down, and smash into the nucleus—all within less than a billionth of a second.

The answer to this conundrum would come by assuming or imposing a new kind of stability and exploring the baffling consequences. As discussed in the next chapter, the solar system model of the atom was replaced by one based on quantum physics. And Einstein showed that the energy produced in nuclear reactions involves an even more radical kind of transmutation in which matter converts to energy.

Box 3.2

The Shape-Shifting Force

While working at McGill, Rutherford categorized radioactive decay into three types: alpha, beta, and gamma. We now know that alphas are the same as helium atoms stripped of their two electrons, betas are high-energy electrons similar to cathode rays, and gammas are a high-energy form of electromagnetic radiation.

The most puzzling of these was beta decay. At first physicists assumed that the emitted electrons came from the cloud that surrounds the nucleus, but the fact that the atom did not become electrically charged after beta decay suggested a different source. Instead, it seemed that an entirely new electron was somehow created by the nucleus.

Furthermore, physicists had known since 1911 that the energy carried by the electron came in a continuous spectrum and was often less than the energy that was known to be produced from the decay. This appeared to contradict the principle of energy conservation. In 1930, Wolfgang Pauli proposed that another particle was being emitted at the same time as the electron. When this was taken into account, energy conservation was restored. In 1933, Enrico Fermi gave this particle the Italian name neutrino and suggested that it interacted, very weakly, with the nucleus through a force he dubbed the weak force.[33]

During beta decay, a neutron transforms into a proton and then ejects an electron and an antineutrino (we will discuss the "anti" later). In the 1930s, neutrons and protons were both considered fundamental particles. The weak force was therefore a shape-shifting force that transmuted one into the other—and it represented the final disintegration of the idea of immutable atoms.

Democritus had defined atoms as the eternal and unchanging constituents of matter. Thomson showed that atoms were made of smaller subatomic particles, and so redefined the idea of the smallest particle. Rutherford demonstrated that the atoms were mostly empty space and furthermore could be rearranged to produce new substances. But the discovery of the weak force meant that even the constituents of the atom were susceptible to change.

The idea of eternal and immutable principles did not go away, but they were themselves forced to transmute; instead of applying to particles of matter, they applied to mathematical laws such as energy conservation and enforced symmetries. As seen later, unified theories see particles such as electrons and neutrinos as being different aspects of the same thing, so the apparent transformation from one form to another is nothing more than a rotation in mathematical space.

While the weak force is far weaker than electromagnetism, it plays a key role in the nuclear chain reaction that powers the Sun and is therefore essential for the energy that supports life on our planet. In fiction, one quality of a good story is that the characters do more than just bounce off one another—they also experience real change. The weak force plays a similar alchemical role in the universe.

No Change

The work of Rutherford and his team led quickly to the development of nuclear physics, as well as associated applications such as nuclear power, medical imaging, radiocarbon dating, and nuclear weapons. Nuclear physics would also finally answer the question of what drives change on our planet. The Sun's energy, which warms the planet and is the ultimate source of most of our energy, is produced by a nuclear fusion reaction in which hydrogen atoms fuse to form helium. Nuclear decay reactions also take place in the Earth's core. While physics has long been based on the idea of stability, many of the most interesting properties of matter—and the ones which support life—involve the processes of change, transmutation, and decay.

The principle of energy conservation remains one of the most useful tools in physics. The physicist Richard Feynman noted that it is "a most abstract idea, because it is a mathematical principle; it says that there is a numerical quantity which does not change when something happens. It is not a description of a mechanism, or anything concrete; it is just a strange fact that we can calculate some number and when we finish watching nature go through her tricks and calculate the number again, it is the same."[34] Experiments at modern colliders like the LHC involve transmutation of the colliding particles into exotic forms of matter, but physicists still use energy conservation to analyze the results. For example, by adding up the energy of particles detected after the collision and comparing with the initial energy of the colliding particles, it is possible to infer whether any other particles have escaped detection (which is also how the neutrino was discovered).[35] Conservation principles are therefore like accountants working at a dynamic and enterprising firm—they can tell us what the constraints are, but say nothing about the real source of creativity and change.

As John D. Barrow noted, all physical laws similarly boil down to statements that "some entity does not change."[36] So what about physical laws themselves—are they also immune to change? Do they hold everywhere in the universe and over all time periods? How confident can we be in applying them to situations such as the first moments of the universe?

Plato believed that mathematical laws exist separately from everyday reality as perfect and eternal Forms. Contemplating them was the highest aim of the student of philosophy. As a character tells Socrates in the dialogue *Symposium*, "Then suddenly he will see a beauty of a breathtaking nature, Socrates, the beauty which is the justification of all his efforts so far. It is eternal, neither coming to be nor passing away, neither increasing nor decreasing. Moreover it is not beautiful in part, and ugly in part, nor is it beautiful at one time, and not at another. . . . It exists for all time, by itself and with itself, unique . . ."[37]

Similarly, many physicists believe that mathematical laws—including the laws of physics—have an independent existence, which is there for us to discover. As Einstein wrote, "I am convinced that we can discover by means of purely mathematical constructions the concepts and the laws connecting them with each other, which furnish the key to the understanding of natural phenomena. . . . In a certain sense, therefore, I hold it true that pure thought can grasp reality, as the ancients dreamed."[38] According to Anthony Zee, a physicist does not invent or create, but "merely discovers a theory which, with its myriad mathematical interconnections, has existed for all time."[39] The physicist Roger Penrose, writing of the "remarkable interrelations between truth and beauty," notes that "aesthetic criteria are fundamental to the development of mathematical ideas for their own sake, providing both the drive towards discovery and a powerful guide to truth. I would even surmise that an important element in the mathematician's common conviction that an external Platonic world actually has an existence independent of ourselves comes from the extraordinarily unexpected hidden beauty that the ideas themselves so frequently reveal."[40]

The belief in stability is therefore tied up inextricably with ancient ideas about aesthetics. A perfect system is one that does not change (for by changing, it could only become less perfect). Physicists seek out entities that are static, in part because it is a very successful approach—as evidenced by the conservation of energy—but also in part because it satisfies a Pythagorean yearning for order and stability. As with the desire for harmony and integrity, this has played an important part in the history of science, but there is a danger that it can become an end in itself and lead to a kind of hubris. As will be discussed in Chapter 7, physicists have extended their models to the birth of the universe on the rather questionable assumption that the laws we observe today were also valid at the time of the big bang.

An alternative to the neo-Platonist view is that mathematical laws, rather than being perfect descriptions of reality, are better described as useful formulae which capture some aspect of a more complex process. After all, for a few brief years at the start of the twentieth century, physicists had what they believed was a more or less stable and consistent model of reality. It was just a question of resolving a few conflicts and addressing some gaps. As Lord Kelvin noted at the 1900 Royal Institution Lecture, "The beauty and clearness of the dynamical theory . . . is at present obscured by two clouds."[41] One involved the question of how objects such as the Earth moved through the "luminiferous ether." The other cloud was what seemed to be a technical issue involving the properties of thermal radiation.

That was when things went nonlinear.

II

COMPLICATION

4

The Crooked Universe

O blinding hour, O holy, terrible day,
When first the shaft into his vision shone
Of light anatomized! Euclid alone
Has looked on Beauty bare.

Edna St. Vincent Millay,
"Euclid Alone Has Looked on Beauty Bare"

You may object that by speaking of simplicity and beauty I am introducing aesthetic criteria of truth, and I frankly admit that I am strongly attracted by the simplicity and beauty of mathematical schemes which nature presents us. You must have felt this too: the almost frightening simplicity and wholeness of the relationship, which nature suddenly spreads out before us.

Werner Heisenberg, letter to Albert Einstein

The track of writing is straight and crooked.

Heraclitus, Fragment 59

Our usual ideas about space and time are based on straight lines and Euclidean geometry. Yet, as Einstein showed, space and time are not straight but are bent by gravity. And it seems that things do not move in straight lines at the atomic level either—instead, they make sudden jumps from one state to another. This chapter shows how relativity and quantum theory together have affected the aesthetics of science and art—and how the inherent nonlinearity of the universe still confounds our attempts to model and understand it.

In the same way that we think of matter as being made up of indivisible fundamental particles, so mathematics is based on a number of axioms—statements which form the foundations of the mathematical edifice and cannot be broken down or derived from other statements. The first complete attempt to codify these axioms was made around 300 BC by Euclid, who was a mathematician at the university and great library in the city of Alexandria.

Euclid starts his *Elements* with a small number of axioms and definitions which describe apparently uncontroversial properties of points, lines, and circles. For example, the first of five axioms proposes that it is possible to draw a straight line between any two points. From these minimalist beginnings, Euclid then goes on to derive the bulk of mathematical knowledge as it then existed, such as the result that there are only five perfect solids.

Euclid's book has been described as the most successful textbook of all time, and has been printed in over a thousand editions. It was an integral part of the quadrivium curriculum for centuries, and continued to be an educational staple well into the twentieth century. Einstein called his copy, which he acquired as a boy, his "holy geometry book."[1] Its logical approach—start with a few basic assumptions and build out from those in a linear fashion—has long inspired scientists and philosophers. It has served as a model for texts ranging from philosopher Bertrand Russell's *Elements of Ethics* to nineteenth-century economist Léon Walras's *Elements of Pure Economics.*

It is also an exemplar of what Russell called the "austere beauty" of mathematics. Statements are proved with no attempt to explain the insight or reasoning which inspired them: as the great German mathematician Carl Friedrich Gauss pointed out, when one has constructed a fine building, the scaffolding should no longer be visible. Because they are proofs, the results hold in all places for all time.[2] Like Plato's Forms, they seem to have an eternal, immutable existence of their own.

While most of Euclid's axioms seem intuitively obvious, the last one—known as the parallel postulate because it implies that parallel lines exist—is a little more tricky, and also more wordy: "if a straight line falling on two straight lines makes the interior angles on the same side less than two right angles, the two straight lines, if produced indefinitely, meet on that side on which are the angles less than the two right angles."[3] This can be expressed a number of different ways, and it turns out to be equivalent to saying that the Pythagorean theorem about right triangles (discussed in Figure 4.1 below) is true.

For centuries, mathematicians attempted to show that the fifth axiom was just a consequence of the other four and therefore disposable, but they found it impossible to do so. In the early eighteenth century, the Jesuit mathematician Giovanni Saccheri adopted the *reductio ad absurdum* approach: he assumed that the fifth axiom was false and tried to prove that this led to a contradiction.

Saccheri spent a major part of his career on this quest, but finally had to admit that—despite being "repugnant to the nature of the straight line"—his assumption was not inherently inconsistent.[4] In the process, he came up with many interesting theorems which would hold true in an imaginary world, or worlds, where Euclid's fifth axiom was false.

By accident, he had discovered non-Euclidean geometry. And those imaginary worlds would include our own.

Post-Pythagorean Geometry

Geometry, which is named from the Greek for earth (gaia) and means of measurement (metron), was originally used as a method to measure out land. One of its founders was Pythagoras, who was credited with many of its theorems. The rows of the Pythagorean tetractys, which encoded the ratios of musical harmony as discussed in Chapter 1, also represented the structure of physical space. The number one corresponded to a single point. The number two corresponded to a line, which could be defined by joining two points. Three represented a plane figure, such as a triangle; and four represented a solid body, such as a tetrahedron with its four corners.

This linear picture of space, in which dimensions are added independently, corresponds to our intuitive picture of the cosmos. We still measure out land using square meters and Euclidean geometry. However, there are other possibilities.

For example, in Euclidean geometry, the sum of the angles of a triangle adds up to 180 degrees. But that only works for flat triangles; it fails when the surface is curved—as on a globe. (In fact it is possible to have triangles where all of the angles are right angles. If from a point near the north pole you walk in a straight line to the pole, turn right, walk the same distance, turn right, and keep walking until you are back where you started, then you will have described a triangle in which each angle is 90 degrees, for a total of 270 degrees.) So Euclidean geometry works when measuring out most pieces of real estate, but it works less well for estimating the area of the Pacific Ocean.

In the early nineteenth century, mathematicians including Gauss started to wonder whether three-dimensional space could also be distorted in a similar way. Gauss kept his research quiet for over two decades, fearful of the controversy it might cause, but in the 1820s other mathematicians came out with their own versions of non-Euclidean geometries. These developments came as a huge shock to philosophers. Instead of there being a single geometry—the one defined by Euclid—there was suddenly a plurality of the things. In place of clarity and uniqueness, there was now confusion.

In Euclidean space, distances are measured in a straightforward way. For example, if we want to compute the diagonal of a room, we just measure each side and use the Pythagorean theorem to obtain it. In non-Euclidean geometry, however, the Pythagorean theorem does not hold and metrics can be considerably more complicated.

Research into the area continued for the rest of the century but found little in the way of practical applications—until a technical expert (third class) at the Swiss federal patent office leaned back on his office chair and felt himself about to tip . . .

That Falling Feeling

"I was sitting in a chair in the patent office at Bern when all of a sudden a thought occurred to me: 'If a person falls freely he will not feel his own weight.' I was startled. This simple thought made a deep impression on me. It impelled me towards a new theory of gravitation."[5]

Einstein may not actually have been tipping back in his seat in 1907 when what he would describe as the happiest thought of his life popped into his mind. But the idea of weightlessness, familiar to astronauts, parachutists, and anyone who has leaned back a little too far in their chair while daydreaming at the office, would lead to the biggest aesthetic shift in science since Pythagoras walked past a foundry and wondered if music was based on numbers.

Einstein's theory of relativity was developed in two stages. The theory of special relativity, which he published in 1905, showed how space and time were linked, applied to all forces except gravity, and was relevant for topics such as radiation and elementary particles. General relativity was Einstein's new theory of gravitation, and it took another decade to develop. Both of these theories introduced new ways of warping geometry.

Special relativity grew out of Einstein's attempt to resolve a contradiction in classical physics. On the one hand, the laws of physics should be the same for all observers who are moving in uniform motion. This principle, known as Galilean relativity, was proposed by Galileo when he was arguing for the Copernican system. He described the following thought experiment:

> Shut yourself up with some friend in the main cabin below decks on some large ship, and have with you there some flies, butterflies, and other small flying animals. Have a large bowl of water with some fish in it; hang up a bottle that empties drop by drop into a wide vessel beneath it. With the ship standing still, observe carefully how the little animals fly with equal speed to all sides of the cabin. The fish swim indifferently in all directions; the drops fall into the vessel beneath; and, in throwing something to your friend, you need throw it no more strongly in one direction than another, the distances being equal; jumping with your feet together, you pass equal spaces in every direction. When you have observed all these things carefully (though doubtless when the ship is standing still everything must happen in this way), have the ship proceed with any speed you like, so long as the motion is uniform and not fluctuating this way and that. You will discover not the least change in all the effects named, nor could you tell from any of them whether the ship was moving or standing still.[6]

For example, the butterflies don't get tired trying to keep up with the boat. Galileo concluded that, in the same way, we could be moving at high speed on planet Earth as it careened around the Sun but be completely unaware of its motion.

Galilean relativity meant that the laws of physics were invariant for uniform motion. However, Maxwell's equations had shown that the speed of light in a vacuum, denoted *c*, is also invariant. If as the boat departs someone on the dock throws a ball at you, then the ball will appear to be traveling at different speeds—faster for him and slower for you, because you are moving away from it. But if instead he flashes a light, the light beam will appear to be traveling at the same speed for both of you.

This makes no sense unless, as Einstein concluded, space and time are fundamentally linked. In particular, observers moving relative to one another must experience time flowing at a different rate. In his *Principia*, Newton wrote that "Absolute, true, and mathematical time, of itself, and from its own nature flows equably without regard to anything external."[7] But it seemed the truth was more complicated.

The mathematics of this was worked out in an elegant fashion by Einstein's former teacher Hermann Minkowski. Instead of viewing space and time as independent coordinates, Minkowski merged them into a single entity, known as spacetime. As he told a 1908 assembly, "Henceforth space by itself, and time by itself, are doomed to fade away into mere shadows, and only a kind of union of the two will preserve an independent reality."[8]

A feature of this spacetime was that its geometry was non-Euclidean. In particular, the distance between any two events involved not just their positions in space, but also their times. Another feature was that only relative position and motion mattered, so there was no single privileged viewpoint—and therefore no need for an Aristotelian ether which imposed a structure on space. This was convenient, since recent experiments by Albert Michelson and Edward Morley had shown that light moved at the same speed in every direction. If ether existed then it would have to be in motion, since the Earth was moving around the Sun. One would therefore expect to find the speed of light to depend on direction.

The theory also held something of a bombshell, for it implied that mass and energy could be transferred into one another. As an object—say a proton at the LHC—is accelerated towards the speed of light, the energy it gains goes not into its speed (which is limited by c), but into its mass, so it effectively becomes heavier. The process can also go in the other direction. When the proton is not moving it has an associated rest mass, which in principle can be converted into energy. The relationship between energy E and mass m is given by what, along with Maxwell's equations, is considered one of the most beautiful equations in physics: $E = mc^2$. Space, time, matter, and energy were thus unified in a single theory.

Einstein's famous equation immediately explained the source of energy in nuclear reactions, such as the radioactive decay studied by the Curies. Some of the atom's mass was converted into energy. As he remarked, "We are led to the more general conclusion that the mass of a body is a measure of its energy-content. . . . It is not impossible that with bodies whose energy-content is variable to a high degree (e.g. with radium salts) the theory might be put successfully to the test."[9]

In fact, the special theory of relativity has been extensively tested in many contexts. For example, the idea that the flow of time depends on motion was tested in 1971 by comparing two highly accurate clocks, one on the ground and the other on an aircraft. The time difference, measured in millionths of seconds, that was recorded after a long flight was consistent with predictions using Einstein's equations.[10] The idea that mass can be transformed into energy was graphically demonstrated by the development of the atom bomb.

All Askew

Einstein originally wanted to name his theory *Invariantentheorie*, or invariant theory, because the point was that the speed of light was not relative, but an unchanging constant. But relativity caught on. The general theory of relativity, which extended special relativity to include gravity, was devised by Einstein as a response to another kind of invariance: the equivalence principle. As mentioned in Chapter 3, this says that the inertial mass which resists motion is the same as (or, more strictly, is proportional to) the mass which experiences the gravitational force. This is why Galileo found that an object's acceleration due to gravity is independent of its mass. Suppose that two balls are dropped from the Leaning Tower of Pisa, one with half the mass of the other. Then the larger ball will experience twice the gravitational force, as per Newton's law of gravity, but it will also have twice the inertia, so the net acceleration will be the same.

For other forces, this isn't true. The electrical force experienced by a charged object in an electrical field depends on the size and sign (positive or negative) of the charge, but this quantity has nothing to do with the object's mass. Gravity is therefore something of a special case. Einstein realized that this could be explained by redefining an inertial system as one that is in free fall.

Everywhere we go on Earth, we are subject to the law of gravity. The only way to escape the stuff is to go to outer space. But we can replicate that gravity-free feeling to a degree in rollercoaster rides when they plunge suddenly down, or in thrill-seeking activities like skydiving or bungee jumping. The experience is terrifying mainly because our bodies suspect that it will suddenly come to a halt.

A falling object can therefore be viewed as one that has in a way freed itself from experiencing force. On the other hand, it still accelerates. Again, this seemed to lead to a contradiction, because as Galileo and Newton had shown, acceleration is usually the result of force. Einstein realized he could resolve the paradox by assuming that gravity distorts spacetime in a non-Euclidean way. Inertial paths were no longer straight lines in Euclidean space—instead they were paths through curved spacetime. In Aristotelian physics, objects want to stay still. In Newtonian physics, they want to move at a constant velocity. But under Einstein, objects just want to fall down.

In 1915 he published the Einstein field equations, which show how the local curvature of a region of spacetime relates to the mass and energy it contains.[11] We experience this curvature in spacetime as a force. As he wrote to his friend Heinrich Zangger, "The theory is beautiful beyond comparison."[12] Unlike special relativity, the conclusions of general relativity were more subtle and harder to test. One of its predictions, though, was that a massive object like the Sun would bend light beams as they passed near it (though it is more correct to say that spacetime would be bending). In 1919, two teams of astronomers confirmed this by measuring slight changes in the positions of stars near the Sun during that year's total eclipse (which blocked the Sun's light and therefore made the stars possible to see).[13] If you were to draw a massive right triangle around the Sun, as in Figure 4.1, then the angles would add up to slightly more than 180 degrees and the Pythagorean theorem would not apply.

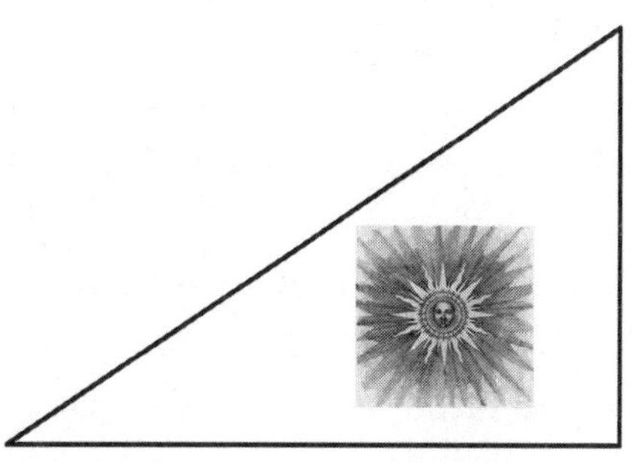

Figure 4.1 For a right triangle, the Pythagorean theorem states that the sum of the squares of the two short sides equals the square of the diagonal. However, if space is distorted by a gravitational mass such as the Sun, as shown, then the angles will add to slightly more than 180 degrees and the Pythagorean theorem will not hold. Source: Adapted from Robert Fludd, *Philosophia sacra et vere christiana seu meteorologia cosmica* (Frankfurt: Francofurti prostat in officina Bryana, 1626).

The newspapers jumped on the story. The *New York Times* headline read, "Lights All Askew In The Heavens—Men Of Science More Or Less Agog Over Results Of Eclipse Observations—Einstein Theory Triumphs." In England the *Times* version was, "Revolution in Science—New Theory of the Universe—Newtonian Ideas Overthrown." As the physicist Paul Dirac later remembered, "[E]veryone was sick and tired of the war. Everyone wanted to forget it. And then relativity came along as a wonderful idea leading to a new domain of thought."[14] Einstein was the first scientific celebrity.

Perfect Universe

Given that mass and energy warp spacetime, a reasonable question to ask is: what shape is the universe? And is it getting bigger, smaller, or neither?

The observable universe contains an estimated 10^{50} tons of visible matter in the form of stars, dust, gas, and so on. (Because the speed of light is finite, there may be distant parts of the universe from which the light has yet to reach us, making them as of yet unobservable.) The universe also contains energy in various forms, including gravitational potential energy. Gravity is a feeble force when compared, say, to the force between electrical charges, but it has infinite extent and exists only as an attraction (contrary to the electric force where unlike charges attract and like charges repel). It therefore adds up.

In Einstein's field equations, the gravitational potential energy for a group of bodies has a negative sign. The reason is that it would take positive work to remove the gravitational force, for example by moving all of the bodies far enough apart to make the gravitational force between them negligible. Estimates show that the net effect of gravity in the universe is to produce about 10^{50} tonnes (in equivalent mass units) of negative energy. The total mass-energy of the universe is therefore close to zero—which implies that the curvature of the universe is also nearly zero. We live in a universe which is approximately flat or Euclidean on the largest scale, but with local distortions caused by mass and energy.

The field equations also contained a term known as the cosmological constant; a force that acted as a kind of pressure to make space expand or contract. The decision to include the term appears to have been primarily aesthetic, for its only purpose was to allow a stable and uniform universe which would not collapse in on itself under the force of gravity. This idea, which other astronomers termed the Perfect Cosmological Principle, was consistent with the ancient idea that the universe was eternal and immutable. Einstein later called the cosmological constant his "greatest blunder" when observations showed that the universe, far from being static, was in fact a dynamic system that was undergoing continuous expansion.[15] However, the concept has recently come back in favor, as discussed in Chapter 8.

When the Perimeter Institute—a physics research institute in Waterloo, Ontario—decided to build a (recently completed) extension, they asked the architects "to provide the optimal environment for the human mind to conceive of the universe."[16] Perhaps they should have considered the patent office at Bern as a model. It must have been unusually conducive to such contemplation, for during his tenure there Einstein also came up with his quantum theory of matter. As with special relativity, this theory had to do with the properties of light—and, once again, it would challenge and transform the aesthetics of science.

Leap of Faith

A time-honored principle of physics was that *Natura non facit saltum* (nature makes no sudden leaps). The world moves in continuous ways and not, as Aristotle put it, in a "series of episodes, like a bad tragedy."[17] In 1900, however, the German physicist Max Planck discovered that from a purely computational point of view it could be useful to think that nature was jumpy.[18]

The discovery came about through a research program at Germany's Imperial Institute of Physics and Technology that was devoted to building a better light bulb. Planck had been analyzing the results of so-called blackbody experiments, which studied how light was emitted from ovens. As a body warms it emits light waves at all wavelengths, with the energy emitted at each wavelength depending on the temperature. (The wavelength is the distance between a wave's consecutive peaks. The frequency of the wave is the time it takes to travel by one wavelength, which for light is equal to the wavelength divided by *c*, the speed of light.) For visible light, the wavelength corresponds to color. For example, when a poker is placed in a fire it glows red, then orange, and eventually white. Classical physics was completely unable to model this effect. Instead, it predicted that the energy would be channeled into short wavelengths of infinite power—the ultraviolet catastrophe, as it became known.

According to this theory, light bulbs would annihilate anyone who turned them on in a sudden burst of radiation. This would not have impressed Planck's sponsor, a local electrical company. However, Planck found that he could model the radiation distribution with a neat formula so long as he assumed that the energy of light could only be transmitted in discrete units or quanta, later called photons. The energy of a photon was equal to its frequency multiplied by a new and very small number, the Planck constant, denoted by *h*.

This approach was incompatible with classical physics because, as Maxwell had convincingly shown, light was a continuous wave. The idea of quanta was therefore viewed—even by Planck—as little more than a computational trick. As he later wrote, he considered it "a purely formal assumption . . . actually I did not think much about it."[19]

At the same time that Planck was studying his ovens, another German physicist named Philipp Lenard was experimenting with the photoelectric effect. When two metal plates in an evacuated jar are connected to a battery, electrons can be made to spark between them when light is shone on the emitting plate. The electrons gain energy from the light and are shaken free from the metal surface. According to the classical wave theory of light, the energy of the emitted electrons should have depended on the intensity (i.e., brightness) of the light source. But in practice, Lenard found that what counted was the frequency of the light. High-frequency blue light created a bigger spark than low-frequency red light.

In another 1905 paper, Einstein argued that the photoelectric effect could be explained using Planck's quanta. He assumed that individual light photons interact with atoms on a one-by-one basis. To knock electrons off their metallic perch, a certain minimum threshold of energy is required. According to Planck's formula, the energy of the photons depends only on their frequency. If the frequency is too low, then no matter how many are sent—that is, no matter how bright the light source—they still have no effect, because as individuals the photons are too weak.

Einstein saw the photons as real particle-like entities rather than a mathematical abstraction. As he wrote, "Energy, during the propagation of a ray of light, is not continuously distributed over steadily increasing spaces, but it consists of a finite number of energy quanta localized at points in space, moving without dividing and capable of being absorbed or generated only as entities." But he still gave his paper the careful title, "On an heuristic viewpoint concerning the nature of light."[20] Like Planck, or Copernicus, he preferred to frame his photoelectric formula as a computational device.

The vast majority of physicists ignored quantum theory or didn't take it seriously. The American physicist Robert Millikan even spent ten years trying to test and disprove it; he could find no fault, but maintained his belief that "the physical theory upon which the equation is based is totally untenable."[21] But the theory would soon find a useful role by helping to make sense of the atom.

The Bohr Atom

In 1912, the Danish physicist Niels Bohr was working in Manchester on Rutherford's solar system model of the atom, trying to find a solution to the model's instability. One problem, as mentioned in Chapter 3, had to do with radiation: according to classical theory, the electrons that circulated like planets around the nucleus would quickly radiate away their energy. Furthermore, electrons would be free to have different radii and energies, so one hydrogen atom could be quite different from another. Bohr hypothesized that the solution was related to Planck's quanta: if the energy of light was limited to discrete units, then so could be the energy of the electron.[22]

Working with the simplest possible case, the hydrogen atom, he supposed that its single electron was restricted to orbits or shells with certain energy levels. In its ground state, the electron will be in the lowest energy orbit, where it can remain without radiating away any energy. If the atom absorbs some quanta of energy, then the electron may jump to a higher energy level. If it then returns to a lower level, a photon is emitted whose energy is equal to the difference in energy between the two orbits.

In effect, Bohr had solved the problem of instability by simply enforcing stability and seeing where it took him. Evidence that he was on the right track came from the fact that his model could explain another perplexing fact about hydrogen that had long puzzled physicists.

It was known that atoms of different elements emit and absorb light at certain distinct, characteristic wavelengths known as spectral lines—like an instrument with its own particular tuning. This allowed scientists to detect which elements were present. For example, Gustav Kirchoff (1824–87) inferred that the Sun was surrounded with a layer of sodium gas by comparing its spectrum with that of table salt (sodium chloride). He also found the signature of an element that at the time was unknown. It was later identified and given the name helium after the Greek sun-god, Helios.

Again, this discrete spectrum didn't sit well with classical physics, which predicted a continuous spectrum of wavelengths instead. In 1885, a Swiss school mathematics teacher called Johannes Balmer discovered a neat and simple formula, based on numerology rather than physics, which seemed to perfectly fit the data for hydrogen. The formula seemed completely mystifying until Bohr showed that it just reflected the different possibilities of orbit transitions in his model of the atom—for example, from level two up to level four, or level three to level one.

With this success behind him, Bohr set about applying his technique to more complicated atoms. In his adaptation of Rutherford's model, the positively charged nucleus was surrounded by circular shells—rather like the crystalline shells that Aristotle had imagined around the Earth—in which electrons orbited. Each shell could only hold a certain number of electrons. The first shell admitted only two electrons, but the next could hold up to eight. If an element had a full outer layer, it was chemically stable, but otherwise it would tend to react. When elements combined to form compounds, they would effectively pool or share the electrons in their outer layers, which would make them more stable.

Bohr's model therefore elegantly explained the chemical reactivity of elements and the structure of the periodic table. Helium is chemically stable—and took so long to be identified by chemists—because it has two electrons and a complete outer layer. The next inert chemical is neon, atomic number ten, with two electrons in the first layer and eight electrons in the second. For the first eighteen elements in the table, the chemical properties tend to repeat in atomic increments of eight—as in John Newlands's chemical law of octaves—because that corresponds to the maximum number of electrons in the outer layer.

Quantum Addresses

The idea behind Bohr's model was certainly simple and elegant; according to Einstein, it represented "the highest form of musicality in the sphere of thought."[23] However, it soon ran into complications. The predictions for spectral lines didn't quite agree with the data, even for hydrogen. And when a substance was placed in a magnetic field, the spectral lines were split—a property which came to be named the Zeeman effect, named for the Dutch physicist who discovered it, Pieter Zeeman.

Progress was made by the theoretician Arnold Sommerfeld (1868–1951), who, playing Kepler to Bohr's Copernicus, suggested that the electron orbits might be ellipses instead of circles. This implied that the energy associated with the electron was now specified by two quantum numbers (the size of a circle is specified by one number, the diameter; but an ellipse requires two numbers, which correspond to the largest and smallest dimensions). Another factor was that a rotating electron produces a magnetic field, and its direction will be affected when an external field is applied. If the electron's orientation, like everything else in the quantum atom, was limited to the discrete values specified by yet another quantum number, then this could explain the Zeeman effect. Sommerfeld compared his work to the Pythagorean belief in numerical harmonies, writing that, "What we are listening to nowadays in the language of spectra, is a genuine atomic music of the spheres, a richly proportioned symphony, an order and harmony emerging out of diversity."[24]

This meant that each electron orbit now had three quantum numbers—two specifying the size and shape of the elliptical orbit, plus one for its orientation. Under closer examination, though, an even finer splitting of spectral lines appeared. It was quickly named the anomalous Zeeman effect. This was starting to annoy people, including Sommerfeld's former student Wolfgang Pauli. When a colleague told him he looked a bit down, Pauli responded, "How can one look happy when he is thinking about the anomalous Zeeman effect?"[25] Pauli proposed yet another quantum number, called spin, which was analogous to the direction in which the electron spins around its own axis. The spin could take two values, denoted up and down.

Again, this seemed like a step in the wrong direction. The model had so many parameters that it had lost any pretense of simplicity or elegance. But Pauli then realized that the extra quantum number could be used to answer a conundrum. Physicists had wondered why electrons didn't just drop down to the lowest energy state, but instead were stratified into Bohr's shells. Pauli proposed in 1925 that the four quantum numbers acted as an address for each electron, like a telephone number. Two electrons could never share the same quantum address.

For example, the two electrons in a helium atom both have circular orbits, so their first three quantum numbers are the same, but they have opposite spin. The Pauli exclusion principle, as it became known, therefore explained the structure of matter. The reason atoms did not collapse in on themselves was that electrons lived at separate addresses.[26]

It was later found that the exclusion principle applies only to the so-called fermions, which include the basic constituents of the atom such as the electron and proton. Bosons, which are responsible for force transmissions and include photons, are quite happy to share the same space.[27] Bosons and fermions are the yin and yang of the universe—the latter give matter its solidity and rigidity while the former hold it together.

Box 4.1

Odd and Even

For the Pythagoreans, the number one was not really a number: it was a principle, the monad, which represented the original state of the cosmos. The dyad, two, was the principle of duality, associated with division and discord. The numbers three, four, and so on were made up from these two principles. Even numbers, which contain the dyad, were considered to be feminine, while odd numbers were masculine. Together, the even and odd numbers gave the universe its properties.

It turns out that, in one sense at least, the Pythagoreans were right. As mentioned above, we now know that matter is divided into bosons such as photons, which can happily share the same space, and fermions such as electrons, which are individualistic and obey the Pauli exclusion principle—like modern house-dwellers, they want their own room. A curious property of the fermions is that they have non-integer spins of sizes equal to an odd number multiplied by 1/2 (1/2, 3/2, etc.). If you take a spinning ball and rotate it by 360 degrees, then the ball and its direction of spin will be the same as before; but if you do the same to an electron, the spin direction is reversed. You have to rotate it twice, or through 720 degrees, to get the same spin.

Bosons are more conventional and have integer spin, which can be expressed as an even number multiplied by 1/2 (2/2=1, 4/2=2, etc.). The physicist Roger Penrose notes that, as with odd and even numbers, "There is a sense in which 'two fermions make a boson' and 'two bosons make a boson' whereas 'a boson and a fermion make a fermion.'"[28] For example, under certain conditions, two "odd" electrons can team up to form a "Cooper pair" that acts as an "even" boson. This pairing allows electrons to flow through the material with little resistance, the phenomenon known as superconductivity. Thus the Pythagorean division into odd and even, which appears also in the Chinese concept of yin and yang, finds its counterpart in the physical structure of matter.

In quantum theory, the Pauli exclusion principle is understood by the fact that swapping two identical fermions introduces a minus sign in the calculation, while swapping two identical bosons has no effect. The probability of a particular state is determined by summing over these two possibilities; it becomes zero in the case of fermions but is enhanced in the case of bosons.[29]

The Musical Atom

While quantum physics had proved useful in explaining a range of phenomena which eluded the classical approach, there were a number of things that didn't add up. For example, what was the justification for Bohr's postulate that electrons in an atom are limited to shells of particular energy? And if light was a particle, then why did it also exhibit wave-like properties? As Einstein told a German newspaper, "There are therefore now two theories of light, both indispensable, and—as one must admit today in spite of twenty years of tremendous effort on the part of theoretical physicists—without any logical connection."[30]

A first step in answering these questions was supplied by the French Prince Louis de Broglie, who was a student at the Sorbonne. In his 1924 PhD thesis, he suggested that if waves can be particles, then particles can be waves as well.

To classical physicists, the clinching proof that light was a wave had come in 1801 when Thomas Young performed his famous double-slit experiment.[31] Until then, most physicists had followed Newton in believing that light was a stream of particles—or corpuscles, as he called them. Young had tested the hypothesis by passing light from a point source through two slits, labeled A and B in his sketch below. Waves have crests and troughs, so when two waves interfere with one another, there will be points where the crests reinforce one another and other points where a deep trough is formed.

This seemed hard to explain using the particle theory. But in 1909 Geoffrey Taylor used a light source as bright as "a candle burning at a distance slightly exceeding a mile" to show that even when individual

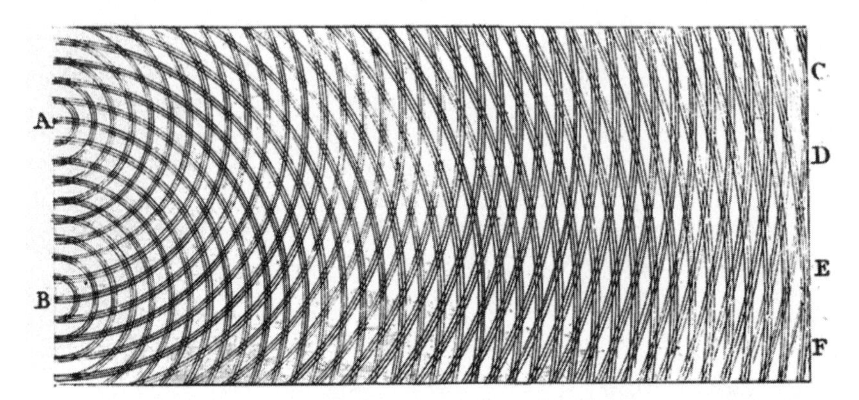

Figure 4.2 Thomas Young's sketch of interference patterns for light waves. Source: Reprinted from Thomas Young, *A course of lectures on natural philosophy and the mechanical arts* (London: Taylor and Watson, 1845).

photons are passed through the slit, the interference patterns are still reproduced.[32] The crests correspond to places with a high probability of seeing a photon, while the troughs are places with a low probability. Even though individual photons only pass through a single slit, they somehow react to the presence of the other slit.

The idea that light could behave like a wave and a particle at the same time appeared to be a paradox. But, as de Broglie pointed out, perhaps the phenomenon was not limited to light. "After long reflection in solitude and meditation, I suddenly had the idea," he later wrote, "that the discovery made by Einstein in 1905 should be generalized by extending it to all material particles and notably to electrons."[33]

By combining Einstein's $E=mc^2$, Planck's equation for the energy of a photon, and the equations for a generic wave, de Broglie inferred that a wavelength should be given by Planck's constant h divided by the momentum. Applying the formula to an electron instead of a photon, he found that the circumference of an electron orbit corresponded perfectly to an integer multiple of its wavelength. He imagined the electrons being guided by pilot waves that vibrate in standing waves like a plucked string, with harmonics that correspond to the energy of the orbit.

Pythagoras had shown that musical harmonies are based on simple mathematics. De Broglie had found a similar relationship for electrons.

De Broglie's PhD examining committee didn't know what to make of the theory, so they passed it on to Einstein for his opinion. His response was, "de Broglie has lifted a corner of the great veil."[34]

Wave Equation

Experiments soon confirmed that electrons did have wave-like characteristics, as de Broglie had predicted. Electron beams diffract like waves when they encounter matter, which is the basis for modern electron microscopes. But if electrons could behave like waves, and light could behave like particles, then the whole idea of the Rutherford-Bohr-Sommerfeld atom, with its hybrid of classical particles and quantum formalism, no longer seemed tenable. It was clear that a radically different approach was needed.

The core of the modern theory was forged in just one year—from June 1925 to June 1926—with competing versions being contributed by Werner Heisenberg, Edwin Schrödinger, and Paul Dirac. Schrödinger began his wave-based version by trying to construct an equation for de Broglie's electrons on the basis that "you cannot have waves without a wave equation."[35]

Physicists were used to working with equations for waves, such as sound waves in the air or the two-dimensional waves set up in the surface of a drum when it is struck. After playing around with different possibilities, Schrödinger found a formula which seemed to work.[36] It was linear in the sense that, like a musical note, the wave function could be decomposed into harmonics representing different quantum states, or added together when more than one particle was present. As he wrote, "I am very optimistic about this thing and expect that if I can only . . . solve it, it will be very beautiful."[37]

The mathematics was daunting, but for simple cases such as the hydrogen atom, Schrödinger could show that the quantum state of an electron corresponds to the harmonics in the solutions to the wave equation. The numbers specifying the quantum address of an electron therefore "occur in the same natural way as the integers specifying the number of nodes in a vibrating string."[38] He had discovered the quantum score for the music of the atom.

Schrödinger's waves were hard to solve but easy for physicists to visualize, and his theory appeared to offer a reconciliation between quantum and classical physics. But this still left the question of what the wave actually stood for—was it a kind of pilot wave, as de Broglie had suggested, or a density of electronic charge? Schrödinger's colleague Max Born suggested that the probability of finding an electron in a particular position was specified by the amplitude of the wave squared.[39]

Places where the wave amplitude was highest would correspond to the likely location of the particle. But because the waves tended to leak out through space, an implication was that particles of any kind—not just electrons—could stray from their usual locations. For example, classical physics had trouble explaining something like alpha decay, because the alpha particle should have remained safely attached within the atomic nucleus. Quantum physics allowed for the possibility that the alpha particle could tunnel through and escape the restraining forces.

The Uncertainty Principle

The German physicist Werner Heisenberg—whose alternative theory, known as matrix mechanics, turned out to be equivalent to Schrödinger's—took a strictly positivist approach to quantum physics.[40] According to Heisenberg, there was no point in speculating about what was going on inside the atom. All one had were observations, such as the vapor trail left by a particle as it passed through a cloud chamber. Even here, though, the measurements had an intrinsic uncertainty. "We had always said so glibly that the path of an electron in the cloud chamber could be observed. But perhaps what we really observed was something much less. Perhaps we merely saw a series of discrete and ill-defined spots through which the electron had passed. In fact, all we do see in the cloud chamber are individual water droplets which must certainly be much larger than the electron."[41]

At a quantum level, the wave-like nature of matter meant that the state of a particle could only be expressed in terms of a smeared-out probability wave. The true state of a particle was therefore intrinsically unknowable. Heisenberg quantified this numerically with his uncertainty principle, which stated that the more accurately the position was measured, the greater the uncertainty in momentum, and vice-versa.

The uncertainty principle offered an explanation for a conundrum which had long puzzled physicists, which was the question of why electrons didn't just spiral into the positively charged nucleus and stop. If an electron were to do so, then its position and momentum would both be exactly known, violating the uncertainty principle. Particles can never be completely at rest, but instead buzz around with some minimum degree of energy. Matter is jittery rather than stable.

The uncertainty principle also applies to other pairs of quantities, such as energy and time—accuracy in one creates uncertainty in the other. An implication of energy uncertainty is that a particle might come from nowhere, exist over a certain time period, and then disappear, all without violating conservation of energy. Such so-called virtual particles are undetectable—they live in a kind of unobservable phantom space allowed by the uncertainty principle—but they still have very real effects. Heavier particles require more energy to produce, so they have shorter lifetimes when produced in this way. Virtual photons have zero rest mass and can extend their reach over long ranges.

For Heisenberg, the uncertainty principle represented a fundamental property of matter. The quantum world at the micro level is different from the everyday world we observe at larger scales. It was therefore a mistake to try to comprehend it using classical concepts. Quantum properties such as uncertainty could only be handled using abstract, Platonic mathematics: "the ultimate root of phenomena is not matter but instead mathematical law, symmetry, mathematical form."[42]

The English physicist Paul Dirac went a step further down the road of abstraction when he invented a new, more mathematical version of the quantum theory; this version incorporated Einstein's theory of special relativity and could therefore be applied to particles moving at relativistic speeds. Dirac arrived at his equation based largely on aesthetic grounds—when asked what he thought of it, he replied, "I found it beautiful." His theory formed the basis of quantum electrodynamics, which modeled electromagnetic force as the interaction of virtual photons.[43]

One curious feature of Dirac's equation, when applied to an electron, was that it had two solutions. In the same way that the equation $x^2=1$ has the solutions $x=1$ and $x=-1$, so the electron's equation could work with a positive charge or a negative charge. At first Dirac thought that the positive solution corresponded to a proton. In this case, the electron might be the only fundamental particle, and everything in the universe made of a single building block—"the dream of philosophers," according to Dirac.[44] Unfortunately, the mass of the proton was out by a factor of about 1,800. The entity described by the equation was instead a positive electron, or positron as it became known. Dirac had shown that a consequence of the union of quantum mechanics and relativity was the existence of antimatter (see Box 4.2).

It would have been possible to produce positrons in the laboratory using a radioactive source, but Dirac had no interest in performing these experiments, and it seems that no one else thought of it.[45] In 1932, at Caltech, the physicist Carl Anderson accidentally became the first person to detect a positron when his cloud chamber, which had been set up to capture cosmic rays, showed a particle with charge equal to that of the electron but curving in the opposite direction in a magnetic field.[46] Even then, it took several months for people to realize that this was the particle predicted by Dirac's theory. When they finally did, it seemed to offer a dramatic confirmation that the inscrutable mathematics of quantum theory, which could only be understood at the time by a small elite of scientists, represented a deep truth about nature.

Box 4.2

Antimatter

According to special relativity, different observers can perceive events as happening at different times. However, because nothing can travel faster than the speed of light, a consequence is that they still normally see events as happening in the same order, thus causality is not affected.

At the quantum level, things are different. The uncertainty principle introduces some slack in timing, meaning that observers can disagree over which of two events happened first.

This paradox is avoided in relativistic quantum theory by noting that a particle moving forwards in time is mathematically equivalent to an antiparticle moving backwards in time. Therefore one observer might see an electron being emitted by a nucleus during radioactive decay, while another observer would see the same nucleus absorb an antimatter electron, also known as a positron. Mathematical consistency therefore demands that every particle have an antimatter equivalent.

The process by which particles and antiparticles are produced and absorbed is the subject of quantum field theory, which is a generalization of quantum mechanics. According to this theory, which was developed in the late 1920s by Dirac, Heisenberg, Pauli, and others, the fundamental constituents of nature are not particles, but fields, which act as stores of energy.

An electron, for example, is the manifestation of an electron field. It in turn is surrounded by a cloud of virtual photons which are the manifestation of an electromagnetic field. The force on another particle is created through the exchange of the virtual photons, and calculations show that its strength decreases with the square of the distance. So Coulomb's law of attraction for the electric field emerges from the actions of these virtual particles.

Unfortunately, the calculations also appeared to show that the electron had an infinite amount of energy (the strength of the electric field approaches infinity near the location of the electron). In the late 1940s, it was found that these infinities could be corrected for simply by subtracting them when they arose, according to certain rules, in a process known as renormalization.

With these corrections, the equations gave excellent agreement with experimental observations. But while renormalization gave the right answers, not everyone was convinced. Dirac said that "I might have thought that the new ideas were correct if they had not been so ugly."[47] And gravitation stubbornly resisted similar treatment and continued to give infinite results.

In the 1970s, quantum field theory was extended to include the weak and strong nuclear forces. The Standard Model of elementary particles is a quantum field theory with more particles and still no gravity. This was progress, but in terms of beauty, were they moving backwards or forwards? As with antimatter, that is in the eye of the beholder.

Fuzzy Logic

Quantum physics was constructed mostly by a small group of young men who seemed to get their best ideas while on outdoor adventures. The twenty-four-year-old Heisenberg came up with his matrix mechanics during a solitary climbing trip to the North Sea island of Helgoland. "At first," he wrote in his autobiography *Physics and Beyond*, "I was deeply alarmed.

I had the feeling that, through the surface of atomic phenomena, I was looking at a strangely beautiful interior, and felt almost giddy that I now had to probe this wealth of mathematical structures nature had so generously spread out before me."[48] Erwin Schrödinger famously invented his wave theory while on an illicit hotel visit with his mistress in the Austrian Tyrol. And, while on a 1927 skiing holiday in Norway, Niels Bohr hit upon his principle of complementarity. This principle said that matter was not a wave or a particle, but a wave and a particle. The two views were not exclusive; they were complementary. Instead of either/or, it was both.

Complementarity was similar in spirit to the philosophy of Heraclitus, a contemporary of Pythagoras. His work is known to us only through a number of brief fragments, often based on paradox, which read like tweets from the depths of time: "Just as the river where I step is not the same, and is, so I am as I am not."[49] He is best known for his statements that "everything is in flux" and "good and evil are one." According to W.K.C. Guthrie, Heraclitus saw the world as "a living organism" in which "everything is always moving up and down the path of change, driven thereto by the attacks of its opposites or its own attacks upon them, but all within strict limits."[50] Opposites are not exclusive, but are two sides of the same coin.

The same idea of complementarity appears in the Chinese concepts of yin and yang. When the Danish government bestowed upon him its highest honor, Knight of the Order of the Elephant, Bohr designed a coat of arms based on the yin-yang symbol and a motto in Latin that read "opposites are complementary."

This was very different from traditional Western logic. In *Metaphysics*, Aristotle wrote that "it is impossible for any one to believe the same thing to be and not to be, as some think Heraclitus says."[51] Light is a wave or it isn't—you have to make up your mind.

Complementarity soon became popular with new-age science writers, but it didn't catch on more generally among physicists. As Heisenberg later noted, "it was found that if we wanted to adapt the language to the quantum theoretical mathematical scheme, we would have to change even our Aristotelian logic. That is so disagreeable that nobody wants to do it; it is better to use the words in their limited senses, and when we must go into the details, we just withdraw into the mathematical scheme."[52] Paul Dirac thought that complementarity "always seemed a bit vague.... It wasn't something which you could formulate by an equation."[53] His much-used textbook *Principles of Quantum Mechanics* was famed, according to one biographer, for "the unadorned presentation, the logical construction of the subject from first principles and the complete absence of historical perspective, philosophical niceties and illustrative calculations."[54] Quantum theory was being Euclidified.

Schrödinger's Cat

Bohr and Heisenberg argued for several weeks over the meaning of quantum complementarity, but eventually arrived at a kind of compromise; this became known as the Copenhagen interpretation and was presented to the luminaries of physics at a famous 1927 conference in Solvay. This reconciled Bohr's duality, Heisenberg's positivism, and Born's probabilities by insisting that our knowledge of a system's state is limited to the results of our observations. Prior to observation, a particle's state is given by a wave function, with inherent uncertainty. Only once a measurement is performed does the wave function collapse, in some unspecified way, so that the quantity being measured has a well-defined value. The act of observation therefore plays a central role, similar to that of a croupier who turns over the cards in a casino to reveal their faces.

Born noted that, while the "motion of particles follows probability laws . . . probability itself propagates according to the law of causality."[55] Still, the use of probability seemed to be giving up on the idea of determinism. This was distasteful to many people, including Einstein, who soon distanced himself from his own invention. He believed that what appeared to be random behavior was actually the result of undetected processes or hidden variables that would appear in a more complete theory of matter. "I find the idea quite intolerable," he wrote in 1924, "that an electron exposed to radiation should choose of its own free will, not only its moment to jump off, but also its direction. In that case, I would rather be a cobbler, or even an employee in a gaming house, than a physicist."[56] In 1944 he wrote to Born that, "You believe in the God who plays dice, and I in complete law and order in a world which objectively exists, and which I, in a wildly speculative way, am trying to capture."[57]

Over the years, Einstein devised a long series of thought experiments to test quantum physics, but he never succeeded in his attempt to prove that the theory was wrong or incomplete. Schrödinger too was skeptical about the interpretation of his wave function, and after discussing with Einstein he came up with a famous thought experiment of his own.

"A cat is penned up in a steel chamber, along with the following device (which must be secured against direct interference by the cat): in a Geiger counter, there is a tiny bit of radioactive substance, so small that perhaps in the course of the hour, one of the atoms decays, but also, with equal probability, perhaps none; if it happens, the counter tube discharges, and through a relay releases a hammer that shatters a small flask of hydrocyanic acid. If one has left this entire system to itself for an hour, one would say that the cat still lives if meanwhile no atom has decayed. The wave function of the entire system would express this by having in it the living and dead cat (pardon the expression) mixed or smeared out in equal parts."[58]

According to the Copenhagen interpretation, the cat is in a suspended state of life and death until a physicist comes along to make an observation and collapse the wave function. This obviously seems unsatisfactory (not least to the cat). Why is it that physicists have privileged observer status but cats do not? What counts as a measurement or an observation? What is the process by which the wave function collapses? And how did the universe get started in the first place with no physicist around to make a measurement and collapse its wave function into actuality?

Despite its drawbacks, the Copenhagen interpretation at least seemed to offer a mathematically consistent story. It was vigorously promoted by its supporters, especially Heisenberg and Pauli—to the extent that other interpretations, according to the physicist John Clauser, were "virtually prohibited by the existence of various religious stigmas and social pressures, that taken together, amounted to an evangelical crusade against such thinking."[59]

Einstein remained unconvinced by quantum theory until his death. In 1935 he and two collaborators proposed another (this time animal-free) thought experiment, known as the Einstein-Podolsky-Rosen paradox. This argued that a consequence of the Copenhagen interpretation was that particles become entangled, so that a measurement on one gives information on the other, even though there is no chance for communication to occur between them—this was what Einstein called spooky action at a distance. As we will discuss in Chapter 10, this seemed to be a fatal flaw in the theory—until it was tested in the 1970s, after Einstein's death, and shown to be correct. Not only did particles lack a clear position and momentum, they also lacked any kind of independence.

Today the Copenhagen interpretation—which physicist David Mermin summed up as "shut up and calculate"—is still accepted as literally true by some physicists, though a number of alternatives exist. These include David Bohm's non-local hidden variable theory, which sees particles as real entities that are guided by the wave function in a manner similar to de Broglie's pilot wave; and quantum decoherence, which explains wave function collapse through interaction with the environment rather than an observer.[60] Perhaps the most popular theory, which seems even more bizarre than the Copenhagen interpretation, is the many worlds hypothesis. As with other theories of physics, it is based on "an ideal of unity, simplicity and completeness."[61] According to this theory, the cat in Schrödinger's experiment is both alive and dead at the same time—but in different worlds. This will be discussed further in Chapter 9.

Quantum Aesthetics

Viewed in aesthetic terms, the quantum atom was very different from the original conception of the atom imagined by Democritus. His motivation for proposing the atom was to show that there was something eternal, solid, and indivisible from which all matter was constructed. The aim was to provide peace of mind through materialism—we and everything around us are just composites of atoms, which come together and fall apart in random patterns. Nothing is real except atoms and the void. Free will is an illusion: everything is the result of what Democritus called "chance and necessity."

The quantum atom, in contrast, seemed to be not firm and solid but both blurred and jagged at the same time. Instead of the calming mechanistic determinism of atomism, quantum theory offered only probability. Instead of smooth, continuous change, it said that nature moved in sharp, random jumps. Instead of independence, it offered spooky connections, coupled with a kind of perverse willfulness. And in place of concreteness, it substituted a fuzzy uncertainty. In quantum theory, the meaning of previously precise concepts like position, momentum, and energy seemed to slide away.[62] Democritean chance and necessity were replaced by quantum probabilities and mathematical wavefunctions (as discussed later, a similar manoeuvre is employed today in other fields), but they didn't have quite the same satisfying ring of solidity.

Einstein's theory of relativity was in some respects equally unsettling. Linear causality gave way to a situation where, as physicist John Wheeler put it, "Matter tells space how to curve, and space tells matter how to move."[63] But to many people it held enormous aesthetic appeal in the way that it unified time, space, and gravity in a single framework. Max Born said that, "The theory appeared to me then, and it still does, the greatest feat of human thinking about nature, the most amazing combination of philosophic penetration, physical intuition, and mathematical skill." He called it "a great work of art, to be enjoyed and admired from a distance."[64] Quantum field theory was also an impressive achievement—but by conventional standards of beauty, with its uncertainties and infinities, it was relativity's ugly sister, and rather far from Sommerfeld's hope for an "atomic music of the spheres."[65]

Some of its equations, such as Schrödinger's wave equation or Dirac's equation for the electron, were certainly mathematically elegant. The physicist Frank Wilczek later described Dirac's equation as "achingly beautiful."[66] But it was a rarefied kind of beauty which seemed to exist only at an abstract level of numbers and symbols. Attempts to interpret their meaning led quickly to confusion, and the idea of wave function collapse seemed a clunky artifice.

Together, relativity and quantum theory completely upended the traditional aesthetics of science. It seems ironic that the end result of a long investigation into the atom, based on linear science, inspired by the "rational mechanics" of Newton, and driven by the search for beauty and harmony, should lead to such an ambiguous and contrived-looking model of the universe. The strict adherence to the orthodoxy of the Copenhagen interpretation, which lasted until at least the 1960s, seems in many respects to be an attempt to save appearances; to emphasize the hard computational power that the theory offered and repress the contradictions at its core.

As Einstein wrote in 1928, "The Heisenberg-Bohr tranquilizing philosophy—or religion?—is so delicately contrived that, for the time being, it provides a gentle pillow for the true believer from which he cannot very easily be aroused. So let him lie there."[67] Aldous Huxley, in a 1929 essay, wrote, "It is fear of the labyrinthine flux and complexity of phenomena that has driven men to philosophy, to science, to theology—fear of the complex reality driving them to invent a simpler, more manageable, and, therefore, consoling fiction. . . . With a sigh of relief and a thankful feeling that here at last is their true home, they settle down in their snug metaphysical villa and go to sleep."[68] In 1930, Ludwig Wittgenstein urged that "Man has to awaken to wonder. . . . Science is a way of sending him to sleep again."[69] The calming, somnolent atomism of Democritus had returned in a different form, but the atom pills weren't quite as effective as they used to be.

Surrealist Science

The dislocations and distortions presented by quantum physics and relativity were not out of tune with developments in the arts. Georges Seurat's technique of pointillism, which represented lights as dots of pure color, preceded the discovery of quanta. The Cubists used the idea of a fourth dimension to justify their use of multiple superimposed perspectives. The work of surrealist artists such as Marcel Duchamp, André Breton, and Salvador Dalí also reflected the same loss of causality, spatial distortions, and uncertainty that their scientific contemporaries were grappling with.[70] The atonal music of Alban Berg and Arnold Schoenberg dispensed even with musical harmony—an omission which, after one 1913 performance, led to a riot. In Lawrence Durrell's 1958 novel *Balthazar*, a character says that "the Relativity proposition was directly responsible for abstract painting, atonal music, and formless . . . literature."[71]

Apparently Einstein did not approve of such uses of relativity, and many other physicists feel the same way.[72] The word has indeed been abused by some writers and artists as a means to add a different kind of gravity to their work. But another way to look at this is to say that relativity and quantum physics together were pointing the way to an alternative conception of beauty—along the lines of Bohr's complementarity—which artists were more willing to take seriously than most scientists. According to this new aesthetic, particles of matter are no longer hard, distinct, and independent, but are fuzzy and entangled. Uncertainty, duality, and paradox are to be taken at face value. Old certainties about order, proportion, and harmony are blown away to reveal twisted new dimensions. Science had its brush with surrealism, but settled for the "tranquilizing philosophy" of Copenhagen instead. As the Cubist artist Georges Braque observed, "Art is meant to disturb, science reassures."[73]

Schrödinger wrote that science is based upon the belief that nature is (1) objectifiable and (2) knowable. As the feminist historian of science Evelyn Fox Keller points out, this implies both a radical split between observer and observed, and a "congruence between our scientific minds and the natural world," which is "not unlike Plato's assumption of kinship between mind and form."[74] But if nature resists objectification and retains a hint of mystery, that is hardly incompatible with beauty. The problem occurs when we try to shoehorn it into a mathematical framework. The beauty of quantum physics is not in the model; just like in a surrealist painting, the beauty is found in the glimpse it offers of another world where time twists and reason warps.

We are used to talking about styles of beauty in other areas such as art or architecture. Science, however, is supposed to search for eternal truths. But when we select what is important and interpret our results, then aesthetics takes over. The Copenhagen interpretation is not a great truth—it is more like a great building whose structure and style reflect the beliefs and biases of its architects. It does the job reasonably well—apart from a few leaks, it can handle the elements—but it looks overly austere, and you probably wouldn't want to live there.

Indeed, the Copenhagen interpretation's emphasis on the detached, objective observer—whose mere presence is enough to magically collapse nature's waves—seems almost a comical parody of the scientific ego. As shown in the next chapter, however, the designers of this theory wouldn't be able to remain objective and detached for long. And they would need more than a reassuring philosophy to get them through the night.

5

The Masculine Philosophy

I remembered the line from the Hindu scripture, the Bhagavad-Gita. Vishnu is trying to persuade the Prince that he should do his duty and to impress him takes on his multi-armed form and says, "Now, I am become Death, the destroyer of worlds." I suppose we all thought that one way or another.

J. Robert Oppenheimer,
"J. Robert Oppenheimer on the Trinity Test"

I felt it myself, the glitter of nuclear weapons. It is irresistible if you come to them as a scientist. To feel it is there in your hands to release the energy that fuels the stars; to let it do your bidding to perform these miracles; to lift a million tons of rock into the sky. It is something that gives people a feeling of illimitable power, and it is in some ways responsible for all our troubles—I would say—this technical arrogance that overcomes people when they see what they can do with their minds.

Freeman Dyson, *The Day After Trinity*

As regards the individual nature, woman is defective and misbegotten, for the active power of the male seed tends to the production of a perfect likeness in the masculine sex; while the production of a woman comes from defect in the active power.

Thomas Aquinas, *Summa Theologica*

The relationship between gender and aesthetics is frequently discussed in areas such as art and literature, but what does it tell us about science? Ever since Plato described women as originating from morally defective souls and Aristotle excluded them from his Lyceum, science has been a game dominated by men. This has affected both the kind of questions scientists ask and the way they interpret the answers. Nuclear weapons, atom smashers, and even the concept of reductionist science all reflect a gendered response to the world. This chapter traces the history of gender bias in science, explores the role played by the militarization of science following the Second World War, and shows how these related factors have shaped the scientific aesthetic.

In his *Republic*, Plato presents his vision of a utopian, rationally ordered society, which is ruled over by a philosopher king and an elite guardian class. The power and integrity of the guardian class would be enforced by both selective breeding—no mixing with the commoners—and rigorous education in the core, number-based subjects of arithmetic, geometry, astronomy, and music (harmony, in particular). Arty activities such as drama, painting, sculpture, and music of the wrong kind were closely controlled because they were potentially dangerous.

According to the theory of Forms, all material objects are imperfect versions of eternal Forms. When a carpenter makes a table, it is one step removed from the ideal Form and is subject to change and decay. If an artist makes a painting of that table, then the result is a copy of a copy, even further away from the perfect reality. Contemplating the painting rather than the Form is therefore deceptive—and, because the artist uses superficial tricks to engage one's interest and emotions, the result is distraction and a failure to achieve a deeper wisdom. Listening to music during class using earphones, had they been invented, would also have been discouraged.

Of course, we do not live ourselves in the world of Forms; we can only contemplate it in an abstract way by using tools such as mathematics and logic. The fact that our minds can engage with Forms while our bodies are stuck in the real world introduces a split in Greek philosophy between mind and body, which manifests itself in numerous ways. Anything associated with the eternal Forms is considered superior to things associated with corporeal reality. Reason, for example, is superior to the senses—one can smell a loaf of bread, but one cannot smell the concept of an atom. Objectivity is superior to subjectivity and prose is better than poetry. The highest of the Forms was the Good.

Plato's *Republic* was perhaps the first detailed discussion in the Western tradition of the philosophy of art, and it has been hugely influential on the development of both art and science. For example, even today many scientists continue to see science in terms of a unified hierarchy with mathematics at the top; followed by physics, which applies mathematics to the universe; then chemistry, which applies physics in order to understand the properties of substances; biology, which studies the chemistry of life; psychology, which studies human behavior and emotions; and finally social sciences like sociology, which studies the interactions of humans.[1] The closer you are to numbers, the closer you are to the top. For this reason, economics is often called the queen of the social sciences.

Mary Midgley's list of opposites for the modern scientist, which was included in Chapter 2, reflects the same split of abstract reasoning over transient emotion. An older version of such a list was produced by the Pythagoreans (see Box 5.1). The message of the Enlightenment, according to philosophers Max Horkheimer and Theodor W. Adorno, was that "anything which cannot be resolved into numbers, and ultimately one, is illusion; modern positivism consigns it to poetry. Unity remains the watchword from Parmenides to Russell."[2]

Gender Divide

Since scientists often cite the importance of aesthetics in their work, this raises the question of which *kind* of aesthetics they are referring to. Aesthetic standards are not static or uniform but are influenced by cultural factors. As feminist philosophers have pointed out, traditional ideas about aesthetics in art harbored a gender bias.[3] The male principle was associated with lofty intellectual ideas, while the female principle was grounded in sensation and the body. More bluntly: men did art, women did crafts. In 1985, for example, The Museum of Modern Art in New

Box 5.1

The Pythagorean List of Opposites

Anyone with young children will be aware of the publishing concept known as the book of opposites—picture books in which facing pages compare opposites such as open/closed, light/dark, and so on. In his *Metaphysics*, Aristotle attributes such a list to the Pythagoreans. In their time it was used for much more than educating children, for the Pythagoreans believed it represented the ten organizing principles of the universe.

Limited	Unlimited
Odd	Even
One	Plurality
Right	Left
Male	Female
At Rest	In Motion
Straight	Crooked
Light	Darkness
Square	Oblong
Good	Evil

The Pythagorean list has much in common with the Chinese dualities of yin, which compares with the right column, and yang, which corresponds to the left column. However, a key difference is that, rather than seeing the two sides as balanced aspects of a unified whole, the Pythagoreans explicitly associated the left column with good and the right column with evil. The list is therefore intended to be dualistic rather than complementary. By aligning themselves with the properties in the left side of the list, the Pythagoreans believed that they could approach divinity.

This list can be viewed as a summary of Pythagorean philosophy, and as such has been influential on the course of Western science. In particular, many of the antitheses can be viewed as a kind of aesthetic key. The two sides of the first pair, Limited vs. Unlimited, were the primary constituents which came together through numerical harmony to form the cosmos, as discussed in Chapter 1. One vs. Plurality reflects the desire for simple principles that reduce a plurality of phenomena to one; At Rest vs. In Motion corresponds to the quest for unchanging laws; and Square vs. Oblong is about symmetry. When scientists speak of a beautiful equation, or sing the praises of symmetry and simplicity, they are reflecting an ancient Pythagorean sentiment.

York opened an exhibition titled *An International Survey of Painting and Sculpture*. As the Guerilla Girls artist collective points out, this was billed as an up-to-the-minute summary of the most significant contemporary art in the world—but out of 169 artists, only 13 were women. "That was bad enough, but the curator . . . said any artist who wasn't in the show should rethink 'his' career."[4] The mind/body duality of Greek philosophy has therefore had a direct and lasting parallel in the world of art, which has only started to break down quite recently. The same holds in science.

Since the time of Pythagoras, it has been a historical fact that numbers and rationality have been seen in the West as male pursuits. In Plato's *Timaeus*, women are described as originating from morally defective souls (though the character may have been representing the Pythagorean opinion rather than Plato's own).[5] Aristotle believed that only men are born with a natural capacity for authoritative reason, and he associated women with passivity and nature.[6] Thomas Aquinas later echoed them when he described the "individual nature" of females as "defective and misbegotten" (he added that, "On the other hand, as regards human nature in general, woman is not misbegotten, but is included in nature's intention as directed to the work of generation").[7]

When the Royal Society was founded in 1660, Henry Oldenburg defined its aim as being the construction of a "Masculine Philosophy."[8] Women were only widely admitted to universities starting in the early twentieth century, with physics departments among the last to open their doors.[9] In C.P. Snow's famous 1959 lecture and book *The Two Cultures*, he added in a footnote that "It is one of our major follies that, whatever we say, we don't in reality regard women as suitable for scientific careers. We thus neatly divide our pool of potential talent by two."[10] As author Margaret Wertheim notes, "Female physicists, astronomers and mathematicians are up against more than 2,000 years of convention that has long portrayed these fields as inherently male. Though women are no longer barred from university laboratories and scientific societies, the idea that they are innately less suited to mathematical science is deeply ingrained in our cultural genes."[11]

In the roll call of most histories of science, female appearances are few and far between, which is unsurprising given the institutional barriers they faced.[12] Emmy Noether, who showed that conservation laws are associated with symmetry principles, had no formal academic job and little pay. Perhaps the most successful woman physicist was Marie Curie—but her husband Pierre Curie still saw women as an obstacle in the battle to think higher thoughts, "For in the name of life and nature they seek to lead us back."[13] As Linda Nochlin wrote in her 1988 essay "Why Have There Been No Great Women Artists?," "The miracle is, in fact, that given the overwhelming odds against women, or blacks, that so many of both have managed to achieve so much sheer excellence, in those bailiwicks of white masculine prerogative like science, politics, or the arts."[14]

Even today, the culture of science remains predominately male, and is given to what physicist Lee Smolin calls a "brash, aggressive, and competitive atmosphere, in which theorists vie to respond quickly to new developments . . . and are distrustful of philosophical issues."[15] He also detects a "blatant prejudice" against hiring women.[16] At the 2011 centenary Solvay conference, which included only two women out of sixty-nine physicists, Lisa Randall reported that the "ratio of x to y chromosomes hasn't changed in 100 years since first Solvay conference in 1911."[17] A 2012 report by the American Physical Society noted that in the United States only 17 percent of physics PhDs are earned by women, and that "science, and especially physical science, is seen by many cultures as a primarily male domain."[18]

Indeed, the gender issue has less to do with headcounts or the sex of individual scientists than it does with what Evelyn Fox Keller calls "gender ideology."[19] According to Keller, few creative or intellectual areas other than science "bear so unmistakably the connotation of masculine in the very nature of the activity." This affects, and is affected by, the number of women in science, but they are not the same thing. Part of our philosophical inheritance is a difficulty in talking about complex, fuzzy issues. According to Aristotelian logic, a statement is either true or false. With something like gender, it is always easy to come up with counterexamples that seem to falsify any statement. Some women are very successful in science, but that doesn't mean that science is gender-neutral.

Some of the more challenging and revealing statements about the role of gender in science have come from people who have experienced both sides of the divide. The biologist Ben (formerly Barbara) Barres wrote in a *Nature* article entitled "Does gender matter?" that most scientists are unaware of any gender bias: "It seems that the desire to believe in a meritocracy is so powerful that until a person has experienced sufficient career-harming bias themselves they simply do not believe it exists."[20] The economist Deirdre (formerly Donald) McCloskey has described how her gender change related to her break with the "main (i.e., male) stream in economics."[21]

Male Gaze

While mathematics and the laws of nature have been associated with the Apollonian male principle, nature itself has traditionally been seen as female. In his book *The Masculine Birth of Time*, the seventeenth-century philosopher Francis Bacon described the role of science as being to "conquer and subdue Nature" and "storm and occupy her castles and strongholds."[22] The roles of scientist and nature are similar to those played by the male observer and the female model in traditional nude paintings, with the detached, objective stance of the scientist corresponding to what art critic John Berger called the "male gaze" of the artist.[23]

It is therefore unremarkable that the aesthetics of science—and particularly the "hard" sciences such as physics—have been characterized by a distinctly male feel. For example, feminist psychologists have noted that the classical picture of the atom as hard, indivisible, independent, separate, and so on corresponds very closely to the stereotypically masculine sense of self (this character appears later in the guise of rational economic man).[24] It must have come as a shock to the young, male champions of quantum theory when they discovered that their equations describing the atom were actually soft, fuzzy, and uncertain—in other words, stereotypically female. The Copenhagen interpretation can itself be interpreted as an attempt to shift the discussion back to the safe ground of conventional "male" mathematics.

Today, of course, the tool of choice for interrogating Nature and peeling back the layers of deception is the particle accelerator. In a recent TED Talk, Murray Gell-Mann described how these "powerful machines" are used to penetrate "deeper and deeper into the structure of particles . . .

We believe there is a unified theory underlying all the regularities. Steps toward unification exhibit the simplicity. Symmetry exhibits the simplicity. . . . That will account for why beauty is a successful criterion for selecting the right theory."[25] Anthony Zee wrote that "physicists from Einstein on have been awed by the profound fact that, as we examine Nature on deeper and deeper levels, She appears ever more beautiful."[26]

It still seems strange that science, which prides itself on rationality and fairness, should have been dominated by one sex for so long. How can it be that a woman didn't obtain tenure at Harvard's physics faculty until 1992? Why did it take until 2009 for the "Bank of Sweden Prize in Economic Sciences in Memory of Alfred Nobel" to be awarded to a woman? The rest of this chapter, which takes up the story of atomic physics, argues that a major reason for this apparent anomaly is the close association between science and another area that has usually been led by men: warfare.

As the economist Lourdes Benería noted, "to deal with gender relations you have to incorporate power into the analysis."[27] And what really locked in the dominance of the male, Apollonian principle in science was what happened next in the history of our relationship with the atom. Rather than rush to embrace their feminine side, scientists quickly resumed their traditional approach of divide and conquer—with explosive consequences. The atom was about to show its strength.

Splitting the Atom

At the same time that quantum theorists were attempting to penetrate the atom using the delicate and finely honed tools of their intellects, experimentalists like Rutherford were wondering if they could find their way in using a more direct means. At Manchester, Rutherford had fired alpha particles—his "trusty right hand"—at gold atoms and used the scattering behavior to infer the size of the nucleus, as described in Chapter 3.[28] But he didn't want to bounce off the nucleus; he wanted to break into it—and to do that, he needed a more powerful bullet.

Rutherford, who had returned from Manchester to Cambridge to head the Cavendish laboratory, realized that a solution was available in the new technologies being developed for high-voltage electrical transmission. These would make it possible to generate powerful electrical fields of several million volts, capable of hurling charged particles towards their targets in a supercharged version of J.J. Thomson's cathode ray tube. The result, he predicted in his 1927 presidential address to the Royal Society, would be "a copious supply of atoms and electrons which have an individual energy far transcending that of the alpha and beta particles from radioactive bodies."[29] Thus was launched the research program which would reach its fruition some eighty years later with the Large Hadron Collider.

Rutherford's proposal seemed speculative in a time when half the homes in Britain were still lit by gas and there was no such thing as a national grid. The most that could realistically be produced in the near term was a few hundred thousand volts, far short of Rutherford's target. But his Cavendish colleague John Cockcroft realized that when it came to breaking the atom, power might not be everything.

As the young Soviet physicist George Gamow had explained at a Cavendish seminar in 1928, alpha decay was a quantum process in which the alpha particles tunneled their way out of the atom courtesy of Schrödinger's wave equation, which allowed a non-zero possibility for the particle to escape the nucleus. In particle physics, energy is measured in terms of electronvolts, with 1 megaelectronvolt (MeV) being the energy gained by an electron after acceleration through one million volts. Alpha particles are emitted with an energy of only about 4 MeV, but according to classical physics they would need 30 MeV to escape the atom. But if the uncertainty principle meant that particles could tunnel their way out, then other particles could tunnel their way in. By bombarding a target with enough projectiles, one could expect some of them to find their way into the nucleus.

Cockcroft estimated that a voltage of about 300,000 volts would be enough for protons to penetrate the nucleus of a substance such as lithium, which is the third-largest element. Together with a student called Ernest Walton, he developed a design for a linear accelerator. This was basically a 2.4-meter-long, vertical version of a cathode ray tube, except that it used the heavier protons instead of electrons. The protons (i.e., hydrogen ions) were supplied by a discharge tube that injected them into the accelerating tube. Key to the design was a clever electrical circuit, still in use today, that could boost the existing electrical supply to the required levels.[30]

The linear accelerator took a few years to take shape, much to Rutherford's consternation, and it was not until April 1932 that the machine made its first operational run, albeit at a reduced power of 200,000 volts. Walton placed a lithium target at an angle of 45 degrees to the vertically-mounted accelerating tube. The protons struck the lithium and the products of the reaction sprayed onto an old scintillation screen used as a detector. When Walton crawled into a small, lead-lined observation box to view the detector through a microscope, he was amazed to see the screen completely aglow with alpha scintillations.

Rutherford was delighted by the results. The details of the reaction were easy to determine. The protons, which were hydrogen ions, had an atomic number of one. Lithium's atomic number was three. An alpha particle was a helium ion with an atomic number of two. One proton therefore reacted with one lithium atom to produce two alpha particles. Or, as a local newspaper put it, "Science's greatest discovery . . . the dream of scientists has been realized. The atom has been split."[31]

Rutherford had little time for this hype, but there was a deeper surprise in store. The protons used in the experiment had a power of 200,000 volts when they hit their targets. When the Cavendish team analyzed the energy of the alpha particles using a cloud chamber, they discovered that they were traveling as if propelled by a force of 8 million volts each. The atom hadn't just split—it had blown up like a grenade, sending shrapnel flying away at an explosive speed.

The source of the energy, they realized, came from the transmutation of mass into energy. The mass of the two alpha particles was 2 percent less than the combined mass of the proton and the lithium atom. That missing mass had been converted into energy. It was the first ever demonstration of Einstein's $E=mc^2$.

The Neutron Gun

The year 1932 was extraordinary for the Cavendish laboratory—and for nuclear science in general—in more than one respect, for it also marked the discovery of the neutron. Rutherford had earlier hypothesized the existence of a neutral particle in the nucleus, on the basis that the atomic weight of many elements could not be explained by their protons alone. But if the neutron had no charge, it would be unaffected by electric or magnetic fields—so how could he possibly detect it?

A hint came in August 1930, when Walter Bothe and his assistant Herbert Becker published results from an experiment in which they had used alpha particles from the radioactive material polonium to bombard light metals such as beryllium. Using a Geiger counter, they found that the target material produced powerful rays, which they believed might be gamma rays. Frédéric and Irène Joliot-Curie (Irène was the daughter of Pierre and Marie Curie) followed up their work and found that the rays from the beryllium could cause protons to be ejected from paraffin. The effect was similar to the photoelectric effect, where ordinary light ejects electrons from a surface, with the difference that protons are about 1,800 times more massive than electrons—so the radiation had to be tremendously powerful.

When the experimental physicist James Chadwick at Cavendish heard about these results, he immediately connected them with Rutherford's neutron. Gamma rays were powerful, but they were high-energy photons, so they had no mass. A neutron, in contrast, had the same mass as a proton and so could easily knock it out of position. In other words, it was possible that the rays were actually beams of neutrons.

Chadwick replicated the experiment by building a kind of ray gun. He managed to obtain some of the expensive polonium from Kelly Hospital in Baltimore, in the form of a thin film over a penny-sized disc. He then simply placed the disc next to another disc of beryllium inside an evacuated tube. The polonium emitted its alpha radiation, which hit the beryllium, which produced the mysterious rays, mostly directed away from the polonium. It was like a machine gun that never ran out of ammunition.

Chadwick first aimed his device at a detector and recorded the radiation emitted from a window at the end of the tube. He then placed a two-millimeter sheet of paraffin between the gun and the detector, and confirmed that a spray of protons had been produced. To measure their energy, he tested how far they could penetrate through thin sheets of

Figure 5.1 Chadwick's neutron gun. Source: Science and Society Picture Library

aluminum foil and calculated that the protons had an energy of 5.7 MeV. This was far more energy than could be produced by a single photon, but it was consistent with the idea of a heavy neutral particle.[32]

Chadwick didn't know it at the time, but his simple gun was the first step on the road to a much more powerful weapon. Neutrons carried no electrical charge, so unlike protons they were not repelled or deflected by the positively charged nucleus. They therefore made a far more effective bullet—and, as it turned out, one big enough to blow up a city.

Chadwick later wrote, on realizing the inevitability of the nuclear bomb in 1941, "I had then to start taking sleeping pills. It was the only remedy."[33] He became a lifelong addict. Einstein's equation, which scientists like Born thought of as "a great work of art, to be enjoyed and admired from a distance," turned out to be more than just a pretty face. And science was about to lose its innocence.

How to Build an Atomic Bomb

In 1938, the German scientists Otto Hahn and Fritz Strassman were carrying out experiments in which they directed slow-moving neutrons at uranium. Elements typically come in different versions or isotopes with a varying number of neutrons. The rarest isotope of uranium, known as uranium-235, has a very large nucleus containing 92 protons and 143 neutrons (which sum to give the atomic weight of 235) and is quite unstable. The experimenters had assumed that the extra neutrons would cause a transmutation to a heavier isotope. Instead, they found that the products were lighter than uranium.

Hahn wrote to his colleague Lise Meitner to ask her opinion. Hahn and Meitner had been a close-knit team at the University of Berlin in the 1920s and early 1930s. But Meitner was Jewish and had recently escaped from Nazi Germany to Sweden. In discussion with her nephew Otto Frisch, who was working with Niels Bohr in Copenhagen, she came to the conclusion that the neutron had broken the uranium nucleus into two roughly equal portions—and, as with the Cockcroft-Walton experiment, some of the mass had gone missing.

Fortunately, wrote Frisch, his aunt worked out that the two nuclei formed by the division "would be lighter than the original uranium nucleus by about one-fifth the mass of a proton."[34] This tiny amount of mass was again converted via $E=mc^2$ into a large release of energy. But in this case, the reaction also released two or three free neutrons. These would then collide with any nearby uranium atoms, thus seeding more divisions. If the quantity of material present was above a certain critical mass—which for uranium-235 was a few kilograms, about the size of a grapefruit—the result would be a self-sustaining chain reaction that grew at an exponential rate, in a process which Frisch compared to the "fission" of dividing cells.

Frisch quickly reported the results to Niels Bohr. "What idiots we have all been!" exclaimed Bohr. "Oh but this is wonderful! This is just as it must be!"[35] Word quickly spread of the momentous discovery. Within days, J. Robert Oppenheimer at the University of California, Berkeley, had sketched out the basic plan for a bomb.

Meitner's contribution to the bomb was never acknowledged by Hahn, or by the Nobel committee who awarded Hahn its 1944 prize "for his discovery of the fission of heavy nuclei." She later wrote to a friend, "I found it quite painful that in his interviews he did not say one word about me, or anything about our years of work together. . . . I am part of

Box 5.2

The Bomb in the Basement

Great artists take risks. So do scientists.

In the early sixteenth century, precious metals including silver were discovered in the Czech town of Joachimsthal, near the German border. Silver minted from the area was known as thaler, from a shortened version of the town's name. Over time, this became pronounced as "dollar."

Another mineral found in the area was Pechblende. The name is from *Pech*, German for bad luck, and *Blende*, meaning mineral. In 1789, the chemist Martin Klaproth found that Pechblende contained a metallic substance that he called uranium, after the recently discovered planet Uranus (and the Greek god of the sky). Uranium soon became popular among the local glassmakers for its beautiful fluorescent glow.

In 1942, scientists experimenting with uranium came dangerously close to making the city of Chicago glow. Nothing captures the gung-ho, risk-taking attitude of the nuclear project as much as the top-secret experiment known as Chicago Pile-1.

The experiment was led by the Italian physicist Enrico Fermi, who had left Fascist Italy before the war (his wife Laura was Jewish) and was the leading expert in nuclear chain reactions. In collaboration with Leo Szilard, he set up the experiment in a rackets court under the University of Chicago's football stadium. The apparatus consisted of a large pile of uranium blocks interspersed with graphite blocks to control the flow of neutrons. Fermi's earlier discovery that graphite is the ideal substance to soak up neutrons and therefore moderate the reaction rate gave the Allies a vital edge over the Nazi effort. Graphite control rods could also be inserted to act as a further brake.

A byproduct of the nuclear reactions would be a new element, named plutonium after Pluto, which was more amenable to fission than uranium. It would be used in the "Fat Man" bomb that exploded over the Japanese city of Nagasaki in 1945.

When the pile, which weighed almost 500 tons, was activated on 2 December 1942 by removing the control rods, detectors showed that it immediately started spewing out neutrons. As Laura Fermi wrote in her book *Atoms in the Family*, "The counters stepped up; the pen started its upward rise. It showed no tendency to level off. A chain reaction was taking place in the pile. In the back of everyone's mind was one avoidable question, 'When do we become scared?'"[36] After a few minutes, Fermi ordered that the control rods be reinserted to shut down the experiment. If his calculations had been out, Chicago might have been the site of the world's first nuclear meltdown.

As it was, Fermi and the other scientists and technicians present were exposed to dangerous levels of radiation. Fermi later died of cancer at the young age of fifty-three. In his obituary, the *New York Times* called him the father of the atomic bomb.[37]

the suppressed past."[38] But as the implications of her discovery became clearer, she was happy to distance herself from it.

Big Science

Plato may have considered poems to be dangerous, but science and mathematics have always been easier to sell to the military. Archimedes is said to have devised a heat ray which used polished metal mirrors to focus solar rays onto enemy battleships during the Siege of Syracuse (c. 214–12 BC). Leonardo da Vinci spent much of his time designing elaborate war machines for his patrons. Galileo marketed his new, improved version of the telescope as a tool—not for stargazing by poets, but for navy surveillance. But it was in 1939 that the military application of science began its ascent to an industrial scale.

Albert Einstein had renounced his German citizenship in 1933 and left for the States in order to take up a position at the Institute for Advanced Studies in Princeton, New Jersey. In July 1939, he was enjoying a vacation at his holiday cabin in Long Island, New York, when two refugee Hungarian physicists, Leo Szilard and Eugene Wigner, came to call. They told him that the German army occupying Czechoslovakia had blocked all exports from the uranium mines in Joachimsthal. The obvious implication was that Germany was trying to build a bomb.

Szilard had contacts at the White House, so Einstein agreed to draft a letter to President Roosevelt. In the letter, he warned Roosevelt that it could soon be possible to set up a nuclear chain reaction: "This new phenomenon would also lead to the construction of bombs, and it is conceivable—though much less certain—that extremely powerful bombs of a new type may thus be constructed."[39] He argued that the government should secure its own supply of uranium and speed up experimental work in nuclear chain reactions, perhaps in collaboration with industrial laboratories.

His letter initially had little effect, and it was not until December 6, 1941, that the United States launched a large-scale nuclear project. The Japanese bombed Pearl Harbor the next day, and soon the Manhattan Project, as it was later named, was in full swing. Over the next few years, the project grew to employ over 130,000 people. Contributions also came from Canada and the United Kingdom, with James Chadwick, the sleepless discoverer of the neutron, heading the UK team. The total budget was almost $2 billion, or about $25 billion in today's dollars. The era of big science had arrived.

The Manhattan Project was under the scientific direction of J. Robert Oppenheimer, a complex character who combined intellectual brilliance with a personality that sometimes seemed as unstable as uranium-235. While a student at Cambridge, he confessed to a friend that he had tried to poison his tutor by leaving an apple laced with laboratory chemicals on his desk. No harm came of it, but the university found out, and Oppenheimer avoided expulsion only after his parents promised to send him to a psychiatrist.[40]

The Allies' resolve to build a bomb was hardened by the knowledge that they were in competition with the German team, headed by Werner Heisenberg. Because most of the other practitioners of the "Jewish science" of quantum physics had left the country, Heisenberg's team was at a disadvantage. Under tremendous pressure from the Nazis, Heisenberg failed to come up with a workable design, apparently because his team lacked the requisite engineering and bomb-making expertise. After the war, some said that the German team hadn't really been trying—a claim that Heisenberg neither made nor denied. As befits the author of the uncertainty principle, Heisenberg's true motivations have never been made clear.[41]

One of the most challenging problems for both teams was acquiring enough of the fissile material—either plutonium, which is a byproduct of nuclear reactions, or uranium-235, which is rare and needs to be separated from the almost identical but more abundant isotope uranium-238. The Manhattan Project pursued both methods. One of the first applications of particle accelerator technology was to help separate out uranium-235.

The first atomic bomb test was carried out in the Jornada del Muerto (Journey of Death) desert in New Mexico, at 5:30 on a Monday morning, 16 July 1945. The bomb, known as "the gadget," was mounted on a steel tower sixteen kilometers from the observation camp. Before the test, the scientists and military officers took out a betting pool on what would happen. Predictions varied from absolutely nothing to an explosion equivalent to forty-five kilotons of TNT. A technician got upset when he heard Enrico Fermi taking out side bets on whether the bomb would incinerate the atmosphere and destroy New Mexico, or even all life on the planet.

In the official report, General Thomas Farrell, who was supervising the test, described what he saw:

> The effects could well be called unprecedented, magnificent, beautiful, stupendous and terrifying. No man-made phenomenon of such tremendous power had ever occurred before. The lighting effects beggared description. The whole country was lighted by a searing light with the intensity many times that of the midday sun. It was golden, purple, violet, gray and blue. It lighted every peak, crevasse and ridge of the nearby mountain range with a clarity and beauty that cannot be described but must be seen to be imagined. It was that beauty the great poets dream about but describe most poorly and inadequately. Thirty seconds after the explosion came first, the air blast pressing hard against the people and things, to be followed almost immediately by the strong, sustained, awesome roar which warned of doomsday and made us feel that we puny things were blasphemous to dare tamper with the forces heretofore reserved to The Almighty.[42]

The *New York Times* journalist William Laurence said he felt like he was present at the moment of creation when God said, "Let There be Light." He described how those present "clapped their hands as they leaped from the ground—earth-bound man symbolizing a new birth in freedom—the birth of a new force that for the first time gives man means to free himself from the gravitational pull of the earth that holds him down."[43] The test site director Kenneth Bainbridge put it more bluntly, "Well, we're all sons of bitches now."[44]

That evening, the physicists held a party to celebrate. Three weeks later, a specially modified B-29 bomber called the *Enola Gay* flew over

Hiroshima. People looking up saw something drop out and a parachute open. They thought this was a good sign, because it meant the plane had been shot down.[45] Then there was a flash of light as the uranium named for a sky god exploded.

Fearful Symmetry

It seems impossible to discuss mankind's most horrible invention—or actually, the first in a series of such inventions—in aesthetic terms. Yet it is true that the bomb gave us a glimpse of a kind of grandeur and beauty. As John D. Barrow observes, the images of the nuclear test show something "strangely beautiful yet immensely destructive, its fearful symmetry pregnant with information about the battered atoms at its heart."[46] The bomb didn't tarnish the mathematical beauty of $E=mc^2$ (the world of Forms doesn't pick up on events back on the ground). It made it more powerful—to some, even glamorous.

The development of the atomic bomb also had a profound effect on shaping and masculinizing the aesthetics of science, if only because it further cemented the connection between science and the military. After the bomb was dropped, Einstein reputedly said, "If I knew they were going to do this, I would have become a shoemaker."[47] While a number of Manhattan scientists became involved in the non-proliferation

Figure 5.2 The "fearful symmetry" of the fire ball created by the Trinity test. Source: Los Alamos National Laboratory

movement, the reality is that very few physicists decided to leave the field.[48] Instead, the weapons program soon led to vast increases in recruitment for nuclear and high-energy physicists—almost all of whom were male.[49] Courses on quantum physics suddenly became popular after the war, not because students wanted to explore complex philosophical questions but because the subject held the key to the most powerful weapon ever devised. The hard-edged, male aesthetic of wartime science became institutionalized.

Physics retained its close military ties during and after the Cold War through groups such as the JASON group of scientific advisors (named after the Greek mythological hero of Jason and the Argonauts), and through a variety of military-related projects such as the Apollo space missions.[50] In many respects this collaboration was highly successful. For example, the Defense Advanced Research Projects Agency (DARPA) was founded after the USSR's *Sputnik* launch to help regain the U.S.'s technological edge over the Russians.[51] One of its many spin-offs was a military communications program which became known as the Internet. Military funding was also behind the development of computers, jet engines, satellite navigation, and lasers, which are used in everything from compact disc players to eye surgery.

As discussed in Chapter 8, the military also nurtured the development of other sciences, including biology and economics, which seemed less directly related to warfare. Again, there was a commensurate effect on aesthetics. It is unsurprising that, as the economist Julie A. Nelson observes, mainstream economics is characterized by an emphasis on "detachment, mathematical reasoning, formality, and abstraction," as opposed to "methods associated with connectedness, verbal reasoning, informality, and concrete detail, which are culturally considered feminine—and inferior."[52] New ideas in economics, presented in Chapter 10, subvert this traditional male aesthetic in interesting ways.

Most scientists are uncomfortable discussing the linked roles of power and gender, writes Lee Smolin, because "it forces us to confront the possibility that the organization of science may not be entirely objective and rational."[53] But we can't talk about the aesthetics of science without acknowledging its historical, psychological, and mythological roots, as well as its connection with the real world. This calls into question the generality of the scientific aesthetic (and indeed the existence of a general aesthetic). How does the search for powerful, universal theorems relate to the more straightforward search for power over others and over nature? How does the desire for scientific objectivity relate to the objectification of nature, and of women? Are traits such as symmetry, unity, and universality essential, or even always desirable? Or are they a projection of a particular gendered worldview?

Of course, scientific theories are not like the designs of wristwatches, where some—like those in the clunky aviator style—are designated "for men"; others, such as the Georg Jensen Vivianna watch, are designated "for women"; and others are unisex. However, it is certainly plausible that we have tended to select areas of study—and of speculation—that fit in with lingering cultural prejudices. As Margaret Wertheim points out, while "we have come a long way since Aristotle, Western society continues to expect that women will remain 'grounded' in the physical, the personal, and the domestic."[54] The quest for "cosmic harmonies" that characterizes much of theoretical physics, in contrast, "is a quest for something utterly *disembodied*, something utterly *immaterial*," that appeals, like mainstream economics, to a traditionally masculine aesthetic.[55]

Consider, for example, the idea of a Theory of Everything—a dream which, as will be seen in the next chapter, scientists have been pursuing for millennia. Stephen Hawking famously said that such a theory would be "the ultimate triumph of human reason—for then we should know the mind of God."[56] John D. Barrow notes that members of the general public often send him amateur versions of such theories, and "in my experience they originate without exception from men rather than women."[57] That seems kind of quirky—until you remember that the professionals are nearly all guys as well. As physicist Marcelo Gleiser notes, "it should not be forgotten that the notion of a Theory of Everything is cultural and not based on scientific evidence."[58] The question of how much effort society should devote to this quest cannot therefore be extricated from the issues of gender, power, aesthetics, and our own history.

As the author Graham Farmelo observes, "The beauty of a fundamental theory in physics has several characteristics in common with a great work of art: fundamental simplicity, inevitability, power and grandeur."[59] Einstein said that only mathematics can supply the "supreme purity, clarity, and certainty" that the theoretical physicist demands.[60] The power, clarity, and beauty of the equations, and what General Farrell called the "clarity and beauty" of the nuclear blast, are linked. They are expressions of what psychologist James Hillman called the "exclusive masculinity, clarity, formal beauty" of the Apollo archetype. The bomb was the ultimate Pythagorean/Apollonian weapon, waging war with the clarity, beauty, and power of the Sun; and its blinding flash etched the scientific aesthetic into our consciousness in a way which has yet to fade.[61]

It's a MAD World

In August 1949, the Soviet Union detonated its own nuclear weapon. During the Cold War, the arms race between the nuclear powers was justified by the doctrine of Mutually Assured Destruction, or MAD. This was developed in the U.S. by the Princeton mathematician John von Neumann, together with military strategists at the RAND Corporation (the name stands for "research and development"). According to von Neumann's mathematics of game theory, rational actors can achieve a stable equilibrium if both know that starting a war will lead to the instant annihilation of both sides.

As a mathematical model of human behavior, MAD makes many of the same assumptions as are common in both physics and economics. It assumes that the world is rational, so it has no crazy leaders. It assumes perfect information, so players always know where an attack originates from. It assumes stability, rather than a shifting dynamic with multiple players. And it assumes symmetry, so that all players have comparable power.[62] As folk singer Malvina Reynolds sang about RAND in the 1960s, "They sit and play games about going up in flames/For Counters they use you and me."[63] (Von Neumann served as one of the models for Dr. Strangelove in Stanley Kubrick's 1964 film of the same name.) The MAD assumptions may have appeared non-mad when it was the U.S. sitting across the table from the USSR—and war was averted, though we came close a number of times—but they are less reassuring in a world of asymmetric power, where the enemy might be a small terrorist cell buying nuclear expertise from rogue states.

Science always seems to be cloaked in an aura of inevitability. The power of the atom existed independently of us, so one could argue that the bomb had nothing to do with issues such as aesthetics or gender—it would have been developed eventually, even if the Manhattan Project was run as a feminist cooperative. The apparent inevitability of science is again tied up with aesthetic arguments. As physicist Steven Weinberg notes, "the kind of beauty for which we look is special. . . . the theories we find beautiful are theories which give us a sense that nothing could be changed. Just as, listening to a piano sonata, we feel that one note must follow from the preceding note—and it could not have been any other note—in the theories we are trying to formulate, we are looking for a sense of uniqueness, for a sense that when we understand the final answer, we will see that it could not have been any other way."[64] Einstein's biographer Abraham Pais compared Einstein's theory of gravity to Beethoven's Opus 135, which proclaims "*Muss es sein? Es Muss sein.*" (Must it be? It must be.)[65]

As shown with the Copenhagen interpretation, though, aesthetic considerations—which are often subconscious—affect the way we see and make sense of the world. They also, as Thomas S. Kuhn noted in *The Structure of Scientific Revolutions*, play a key role in selecting what is considered to be "good" science.[66] There is usually more than one paradigm or model that can be used to describe a natural phenomenon, so any approach which appeals to standard ideas of attractiveness is likely to dominate. This is particularly important when confirmatory data is lacking or when all models have poor predictive ability because of the complexity of the system. As will be seen in later chapters, aesthetics can be used to justify theories that are highly speculative or just wrong.

Aesthetic considerations also come to the fore when scientists make decisions about what areas of science deserve the most funding and resources. The course of postwar physics was directed by the scientists who had worked on the bomb. The pivotal event was a meeting at Shelter Island, a secluded location off Long Island, New York, where twenty-four of these scientists—all men and most young—gathered to discuss the foundations of quantum mechanics and set the research agenda for the future. The meeting was "small, closed, and elitist in spirit."[67] The conference attendees, who included Oppenheimer, von Neumann, and Richard Feynman, were treated like celebrities, and were even supplied with motorcycle police escorts when they arrived.[68]

After the meeting, the area which came out best in terms of funding was particle physics. This might seem hard to understand, given that it is occupied with esoteric research which offers little in the way of direct real-world applications and can be understood only by an elite class of highly-educated scientists.[69] As seen in the next chapter, the reason once again was related to our ideas about beauty.

6

Unity

Truth is ever to be found in the simplicity,
and not in the multiplicity and
confusion of things.

Isaac Newton, *Fragments from a Treatise on Revelation*

When judging a scientific theory, his own or another's, he asked himself whether he would have made the universe in this way had he been God. This criterion may at first seem closer to mysticism than to what is usually thought of as science, yet it reveals Einstein's faith in an ultimate simplicity and beauty in the universe. Only a man with a profound religious and artistic conviction that beauty was there, waiting to be discovered, could have constructed theories whose most striking attribute, quite overtopping their spectacular successes, was their beauty.

Banesh Hoffmann, *Albert Einstein: Creator and Rebel*

And someday, we may actually figure out the fundamental unified theory of the particles and forces, what I call the "fundamental law." We may not even be terribly far from it. . . . And when the mathematics is very simple—when in terms of some mathematical notation, you can write the theory in a very brief space, without a lot of complication—that's essentially what we mean by beauty or elegance.

Murray Gell-Mann, *Murray Gell-Mann on Beauty and Truth in Physics*

If science has a great unifying belief, a principle with which all scientists are seemingly united in agreement, it is the statement that science is about unification. A beautiful theory is one which unifies the greatest range of phenomena in the most concise way. Famous examples include Newton's law of gravity, which unified the motion of an apple with that of a planet around the Sun; Maxwell's equations of light, which unified electricity and magnetism; and Einstein's E=mc^2, *which unified energy and mass. In the 1960s, physicists developed the so-called Standard Model, which made sense of the profusion of particles produced by accelerator laboratories but failed to incorporate gravity. Today, the search for a grand unified theory is more intense than ever. But has the desire for unification lost touch with reality?*

In 1905, at the age of twenty-six, the patent clerk Albert Einstein published five papers that rewrote the laws of physics. The first, in March, showed that light came in quanta. The next two, in April and May, convincingly proved for the first time that atoms exist and estimated their size. June brought the theory of relativity; and in September he followed up with a brief note that contained his famous formula, $E=mc^2$. It was as if a whole new theory of the universe had sprung fully formed from his brain. It wasn't big science—it was just big.

Quantum physics was a young man's game, and it was often said that most of its practitioners lost their edge by the age of thirty.[1] Brilliance, it seemed, had a short half-life. Indeed, Einstein's output began to show inevitable signs of decay as he aged. After the war, he spent most of his time trying to devise a single theory that would unite physics at the level of quanta with general relativity. While experimental evidence for both of these separate theories soon accumulated, no one, including Einstein himself, could find a way to make them compatible with each other. General relativity worked well at large scales and electromagnetism worked at small scales, but once you applied general relativity to quanta the equations blew up and became infinite.

For years Einstein struggled without success to "reduce to one formula the explanation of the field of gravity and of the field of electromagnetism."[2] Trying one strategy after another, he showed little interest in new experimental or theoretical results, such as the discovery of the weak and strong nuclear forces which hold the nucleus together.[3] Increasingly isolated for his contrarian stance that "God does not play dice," he became something of an exile from the physics community.

Einstein never gave up on his increasingly quixotic quest. Perhaps he felt that the failure to reconcile his two great inventions somehow invalidated their integrity. Or perhaps he was driven by aesthetics. As the science historian Yehuda Elkana noted, "For Einstein, the concept of simplicity is intimately connected with the concepts of harmony, beauty, and symmetry. There must be one, generalized, simple concept of field or law. . . . Asymmetry is disruptive, unnatural, misleading."[4] Physicist Hermann Bondi wrote of Einstein that "as soon as an equation seemed to him to be ugly, he really rather lost interest in it and could not understand why somebody else was willing to spend much time on it. He was quite convinced that beauty was a guiding principle in the search for important results in theoretical physics."[5] When Einstein's colleagues went into his office after his death in 1955, the blackboard was covered in detailed equations, but none of them added up to the elusive thing that he was searching for: a Theory of Everything.

Einstein's story is just one segment of a project that has fascinated, obsessed, and frustrated scientists and philosophers for thousands of years. Today, scientists are working harder than ever towards their goal of a Theory of Everything, which will reduce the fundamental laws that govern absolutely everything in the universe down to a few simple, sparse, and elegant lines—and preferably in a format that, as Leon Lederman puts it, "will fit easily on the front of a T-shirt."

But can the universe really be reduced to a slogan? Or will it turn out to be even more complicated than a modern electrical appliance, which comes with hundreds of pages of documentation in several languages? Is there an instruction book at all? This chapter charts the history of unification—and asks whether what we actually need is more diversity. As Nietzsche wrote in *Beyond Good and Evil*: "Love of one is a barbarism; for it is exercised at the expense of all others."[6]

Love of One

Einstein's remarkably productive year of 1905 is sometimes called his annus mirabilis, or miracle year. In 1666, Isaac Newton had a similar hot streak. Spending the summer at his mother's farm in order to avoid the plague that was sweeping Cambridge, he not only discovered his law of gravity, but came up with the three laws of motion, co-invented calculus, and decomposed white light using a prism to show that it was made up of all the separate colors.

It still seems amazing that things as varied as gravity, the motion of bodies, and the behavior of light could be amenable to the same kind of mathematical treatment. Nature, it seems, "is very consonant and conformable to her self," as Newton wrote in *Opticks*.[7]

Coupled with our scientific desire for mathematical unification is a belief that the resulting equations should be very simple and elegant. This simplification is usually accomplished by exploiting some type of symmetry in the system. Unity and symmetry are therefore two sides of the same coin. As Anthony Zee puts it, "Physicists dream of a unified description of Nature. Symmetry, in its power to tie together apparently unrelated aspects of physics, is closely linked to the notion of unity."[8] Newton's laws of gravity and motion were based on a symmetry between masses: the Earth exerts a force on the Moon, but the Moon also exerts a force on the Earth. In particular, it pulls on the sea to produce tides. The law of gravity was also symmetric in a geometric sense, since the force produced by a body is the same at all points a given distance away. Kepler was despondent when he found that the planets moved in asymmetric ellipses, but Newton restored symmetry by showing that the underlying equations were symmetric.

In 1861, James Clerk Maxwell wrote out a set of equations which simultaneously described the electric and magnetic forces. This unification led him to the idea that light was a form of electromagnetic wave. His original formulation consisted of eight equations describing the forces in the three different directions that make up three-dimensional space. The equations were later reduced to just four by way of a notation which accounted for the fact that Euclidean space is symmetric—in other words, there is no distinction between up or down, left or right.

The equations can also be expressed in even more compact ways. In quantum physics, it is usual to write the fields in terms of a magnetic and electric potential. The strength of the field can be determined by performing a mathematical operation (differentiation) on the potential. When written in this way, the number of equations reduces to only two. However, this simplicity comes at a cost; the potentials must be defined separately and are not unique—as a result, different formulations can be used to give the same answer, so long as they respect the so-called gauge conditions. In a superficial way, the equations look simpler, but behind the scenes things are getting messy.

Maxwell's equations were developed in the framework of Euclidean geometry. Einstein's 1905 paper on relativity begins by noting that, when they are applied to moving bodies, the equations lead to "asymmetries which do not appear to be inherent in the phenomena."[9] His investigation and correction of these asymmetries—specifically the fact that the speed of light is the same for observers who are moving relative to one another—led to his theory of special relativity and the identification of mass with energy.[10]

The mathematics of special relativity also permitted the simplest and most elegant formulation of Maxwell's equations. It combined the electric and magnetic potential into one single potential which exploits additional symmetries between the two. The result looks like this:

$$dF = 0$$
$$d^*F = J$$

For a physicist, that's beautiful. That's elegant. It is so simple it looks like a tautology—even a necessity. And it definitely fits on a T-shirt. Of course, you have to know what F and J and the other symbols mean, but that's in the small print.[11]

Unlike Maxwell's original equations, this formulation doesn't give you any sense of the fields that you might actually measure. There is no hint, for example, that a point charge produces an electric force on another charge that decreases with the square of the distance (Coulomb's law). The mathematics has moved from the world of observable forces to a higher plane of symmetry and unity.

Given the successes of Newton, Maxwell, and Einstein, it is perhaps unsurprising that science has come to be seen as almost identical to the idea of unification. According to science historian Jacob Bronowski, for example, "Science is a process of creating new concepts which unify our understanding of the world."[12] Mathematician Hermann Weyl wrote, "I am bold enough to believe that the whole of physical phenomena may be derived from one single universal world-law of the greatest mathematical simplicity."[13] Experimental physicist Leon Lederman: "Unification, the search for a simple and all-encompassing theory, is the Holy Grail."[14] Theoretical physicist David Deutsch: "The attainment of a Theory of Everything will be the last great unification."[15] Cosmologist Paul Davies: "All science is a search for unification."[16] Theoretical physicist Anthony Zee: "The ambition of fundamental physics is to replace the multitude of phenomenological laws with a single fundamental law, so as to arrive at a unified description of Nature."[17] Steven Weinberg, who has been called the master builder of the Standard Model: "a single unified theory . . . is what we are working for and what we spend the taxpayers' money for."[18]

Theoretical physicist Lee Smolin humorously notes that "the most cherished goal in physics, as in bad romance novels, is unification. To bring together two things previously understood as different and recognize them as aspects of a single entity—when we can do it—is the biggest thrill in science."[19] Underlying all of this, of course, is the belief that nature itself is unified, both in terms of mathematical laws and the fundamental stuff of which the universe is made. As John D. Barrow notes, "The unity of the Universe is a deep-rooted expectation."[20]

While a sense of unity and wholeness—the feeling of being connected to a larger purpose—is a feature of many (particularly monotheistic) religious and spiritual traditions, in science it is set as an explicit goal to be achieved using mathematics. But where does this bold faith in unity, which does often seem religious in its nature, come from? Why is everyone united about unity? Does it represent a great insight about nature? Does it tell us more about ourselves?

The desire to achieve unification can be seen as the desire to wind the cosmos back to its original, primeval state of unity; to rediscover the Pythagorean monad. But when physicists turned their attention from nuclear chain reactions to atom smashers after the war, they found there was just one problem: too many particles. Squaring the universe turned out to be even more difficult than they thought.

Accelerator Race

Prior to the war, most particle physicists only had access to two types of sources for their particles. One was naturally occurring radiation, such as the beta particles used by Rutherford to probe the structure of the atom, which had energies of only about 4 MeV. The other was far more powerful, but was also much harder to focus or control: outer space. Since the 1910s, it had been known that the Earth was constantly being bombarded from above by highly energetic particles.

Early researchers into radioactivity noticed that particles were sometimes detected even in the absence of any known source of radiation. The Jesuit priest Theodor Wulf thought these were emitted by the Earth, but when he lugged a detector up to the top of the Eiffel tower in 1910 to test his hypothesis, he found instead that the signals got stronger at a higher elevation. The Austrian physicist Victor Hess carried out similar experiments in a balloon and discovered that the radiation at five kilometers in the air was higher still. He concluded that the source of the radiation was outer space and gave it the name "cosmic rays." More recently, astronauts in the Apollo program noticed that they would sometimes be disturbed while trying to sleep by the occasional flash of light as a heavy particle hit the retina of their closed eye.

It was eventually discovered that the cosmic rays were caused mostly by high-energy photons and protons, which collided with the upper atmosphere to create a shower of particles. Most were absorbed by the atmosphere, but some made it through and could be detected using a variety of techniques. One method was to layer a series of a hundred or so photographic emulsions on top of one another and leave the stack on the top of a mountain—like a trap for wild particles. After developing the plates, examination by microscope would reveal a kind of 3D photograph that tracked the progress of particles through the layers. Use of this and other methods such as the cloud chamber led to the discovery of a number of bizarre new particles that didn't fit with the conventional picture of the atom. These included the positron (1932), discussed in Chapter 4, which is the anti-matter version of the electron; the muon (1937), which is like a heavy and unstable relative of the electron; and charged pions and kaons (1947), which have a mass of about two-thirds or less than that of a proton.

While cosmic rays gave tantalizing glimpses of these exotic and short-lived forms of matter, it soon became obvious that physicists needed a steady, controlled supply of particles to get a proper picture of the subatomic world—and that the resolution of the picture depended on the power of the machine. Particle physics, as will be discussed further below, therefore progressed in lockstep with the development of particle accelerators, which we will discuss first.

The first particle accelerator to be designed was that of Cockcroft and Walton (see Chapter 5). This device was called a linear accelerator because it fired the protons in a straight line from a source at one end to the target at the other. An alternative design was the cyclotron, built first by Ernest Lawrence and his team at Berkeley, in which the protons were bent in a circle by a magnetic field. Each time they completed a loop, they were given a push by an alternating electric field. As they accelerated, the radius of their path increased until they were finally ejected.

An advantage of the circular design—at least initially—was its compactness. Lawrence's first model was about ten centimeters in diameter, but by 1939 this had expanded to 1.5 meters. He managed to attain funding for an even more powerful 4.7-meter machine, but the magnet for this design was appropriated by the Manhattan Project for the enrichment of uranium-235.

After the war, the prestige of physicists was at an all-time high and money for high-energy physics was readily available, especially in the United States. Lawrence's annual budget at Berkeley, which had been $85,000 before the war, was cranked up to $3 million. Since larger cyclotrons would have required a ridiculously large magnet, designers switched to a synchrotron design in which a number of magnets were arranged in a circular configuration, like beads on a necklace. As the particles accelerated, the magnetic field was ramped up in tandem to keep them on their course. This allowed the beam to be stored and manipulated or ejected at will.

The 1950s saw adoption of a new unit of measurement, the gigaelectronvolt (GeV), to describe accelerators—with 1 GeV being equal to 1,000 MeV—and the subsequent construction of the 3-GeV Cosmotron at Brookhaven National Laboratory on Long Island in 1952, followed two years later by the 6.2-GeV Bevatron at Berkeley. In 1957, the Soviet Union joined the race with a 10-GeV synchrotron at Dubna. But the accelerator race really accelerated when Russia launched the *Sputnik-1* satellite in October of that year.

As the BBC noted, the rocket that put it into space "might also be capable of carrying a nuclear weapon thousands of miles. The fact that Sputnik is expected to fly over the US seven times a day has also caused unease. There have already been calls for an immediate review of US defenses, given the implications of the technological leap ahead by a political enemy."[21] In response, President Eisenhower ordered increased spending on science education, and in the next four years Congress doubled federal spending on research and development, with much of the funding flowing to particle accelerators. The race to space would be mirrored by a race to the atom.

More powerful accelerators required more powerful detectors. Chief among these in the 1950s was the bubble chamber, which was similar to a cloud chamber but was filled with pressurized liquid hydrogen instead of water vapor. This slowed the transit of particles, created interactions between them, and improved the likelihood of their detection. The high pressure allowed the liquid to be maintained in a superheated phase, well above its usual boiling point at room pressure. Particles passing through it caused it to locally boil along their path, leaving a clear trace that could be photographed from a number of different angles. A uniform magnetic field was applied so that charged particles would curve by an amount which depended on their momentum and charge. Physicists could use this information to piece together which particles were produced following a collision.

Not to be left out, European nations banded together in 1954 to form the European Organization for Nuclear Research, known by its French acronym of CERN. In 1959, their Proton Synchrotron (PS)—which is still being used today to inject protons into the LHC—reached a record beam energy of 28 GeV. The record was soon lost to the Alternating Gradient Synchrotron (AGS) at Brookhaven National Laboratory, which attained 33 GeV in 1960. The name referred to the technique of using magnetic fields to focus the proton beam. This allowed higher energies to be obtained without having protons collide with the wall of the beam tube. It in turn was overtaken in 1967 by the Soviet Union's 70-GeV accelerator at Serpukhov, followed in 1972 by the U.S. National Accelerator Laboratory, which was later renamed Fermilab (see Box 6.1) after Enrico Fermi. CERN responded in 1976 with the 400-GeV Super Proton Synchrotron (SPS).

Box 6.1

Fermilab

The construction of Fermilab was directed by Robert Wilson, who had been a group leader on the Manhattan Project. He came from frontiersman stock and liked to inspect his facility on horseback, wearing a cowboy hat. To symbolize both the area's connection with the prairie and the pioneering spirit of science, he installed a small herd of free-grazing bison who roamed the grounds like cantankerous physicists.

In order to attract scientists to go and work in a cornfield outside Chicago, he decided to be as ambitious as possible, in terms of both beam energy—he proposed 500 GeV—and delivery—the project was completed on time and under budget. An accomplished sculptor, he also placed aesthetics at the core of the project. "I envisaged the Laboratory," he later wrote, "as a utopian place where physicists coming from all parts of the country—and from all countries—would be doing their creative thing in an ambiance of well-functioning and yet beautiful instruments, structures, and surroundings that would reflect the aesthetic magnificence of their discoveries and theories."[22]

The design of the main hall was modeled after the Saint-Pierre Cathedral at Beauvais, France. Wilson noted "a strange similarity between the cathedral and the accelerator: The one structure was intended to reach a soaring height in space; the other is intended to reach a comparable height in energy. Certainly the aesthetic appeal of both structures is primarily technical. In the cathedral we see it in the functionality of the ogival arch construction, the thrust and then the counterthrust so vividly and beautifully expressed, so dramatically used. There is a technological aesthetic in the accelerator, too. There is a spirality of the orbits. There is an electrical thrust and a magnetic counterthrust. Both work in an ever upward surge of focus and function until the ultimate expression is achieved, but this time in the energy of a shining beam of particles."[23]

As with a cathedral, or a Vitruvian temple, aesthetics informed many of the lab's key proportions. Since Wilson had "a propensity for round numbers," the radius of the main ring was chosen to be exactly 1 kilometer. Wanting to make the circle visible, he marked it with a six-meter-wide berm of earth, accented with a canal that carried cooling water for the magnets.[24] Satellite pictures show the earth berm as a perfect circle in the Illinois landscape, like an ancient earthwork.

To determine the ideal height for the main building, Wilson took a helicopter up to plot an "aesthetic factor" for the view of the prairie countryside depending on elevation. He made a number of mathematically inspired pieces of sculpture for the grounds, had significant input on the architectural design of the buildings, and along with resident artist Angela Gonzalez tried to imbue the entire laboratory—down even to the machinery that filled the tunnels—with a kind of beauty. "I have always felt," he wrote, "that science, technology, and art are importantly connected, indeed, science and technology seem to many scholars to have grown out of art. In any case, in designing an accelerator I proceed very much as I do in making a sculpture. I felt that just as a theory is beautiful, so, too, is a scientific instrument—or that it should be. The lines should be graceful, the volumes balanced. I hoped that the chain of accelerators, the experiments, too, and the utilities would all be strongly but simply expressed as objects of intrinsic beauty."[25]

As the rhetorician Joanna Ploeger pointed out, the emphasis on aesthetics also helped to dissociate the project—at least in the eyes of the general public—from weapons research and the increasingly unpopular nuclear program, while at the same time fueling a sense of mystique which helped to insulate it from criticism. Gonzalez's art (see Figure 6.1), for example, "references the magical, even mythical, qualities of the laboratory and its work. . . . There is indeed something magical and unbelievable about the fact that we, as a society, should choose to create and continue to support such an elaborate scientific and technological enterprise—a speculative creation with *no obvious practical purpose*."[26] So, rather like an art gallery, then.

When Wilson stepped down in 1978 to protest a lack of funding, he was replaced by Leon Lederman. The main ring at Fermilab experienced a number of upgrades and remained in use until 2011. Lederman wrote, "One must consider the aesthetic qualities of an accelerator as well as its GeVs and other technical attributes. Thousands of years hence, archaeologists and anthropologists may judge our culture by our accelerators. After all, they are the largest machines our civilization has ever built. . . . Accelerators are our pyramids, our Stonehenge."[27] The 1-kilometer-radius circle of its ring will stand forever as a symbol of our quest for a unified theory.

In addition to proton accelerators, a number of less expensive electron accelerators were also constructed. When electrons traveling near the speed of light are deflected by magnets, they radiate energy in the form of high-energy X-rays that are emitted at a tangent to the path. This so-called synchrotron radiation, which is short-wavelength light, is useful for a number of purposes, such as probing the structure of materials.[28] A drawback is that the energy lost to radiation must be constantly replaced. Since linear accelerators do not need to bend the electrons in a circle, they avoid this problem. The Stanford Linear Accelerator Center (SLAC) was built in a 3-kilometer straight tunnel just west of the Stanford University campus in California. Operational since 1966, it is still the world's largest linear accelerator.

Figure 6.1 Artwork by Angela Gonzalez showing the main building at Fermilab. The sculpture at the bottom, called the Mobius Strip, is one of the site's sculptures designed by Wilson. Source: Fermi National Accelerator Laboratory, Batavia, Illinois

Collider

In these early accelerators, the beam of protons or electrons would be steered into a target, such as a lump of metal, and the products analyzed using detectors. Acceleration did not translate into greatly higher particle speeds—they were already traveling at near the speed of light—but it did mean that there was more energy available for the collision. However, physicists realized that this wasn't the most efficient arrangement. When a moving car hits a car that is at rest, much of the energy goes into displacing the stationary car; but in a head-on collision, all of the energy is absorbed by the cars themselves, causing maximum damage. In the same way, two particles colliding head-to-head would give the greatest bang for the buck. The challenge was to attain a high enough beam density that the rate of collisions would be sufficiently great.

One such collider, built at SLAC and completed in 1972, was the 3-GeV Stanford Positron Electron Asymmetric Rings (SPEAR). As the name suggests, it collided a beam of electrons with a beam of positrons. An advantage of using positrons is that they have the same mass but the opposite charge as electrons, so they bend the opposite way under a magnetic field (while a disadvantage is that antimatter particles such as positrons are harder to produce in high densities). The two beams can therefore circulate in opposite directions in adjacent beam tubes, which merge together in the detector halls. SPEAR was later superseded by the Large Electron Positron (LEP) collider at CERN, which was located in a massive 27-kilometer tunnel straddling the French-Swiss border. At full power, the LEP's electricity consumption was equivalent to 40 percent that of the entire city of Geneva. The machine operated from 1989 to 2000, when it was closed down so that the tunnel could be reused for the LHC.

Machines based on the heavier protons rather than electrons were also constructed on both sides of the Atlantic. At Fermilab, under Leon Lederman, the magnets of the main ring were removed and replaced by powerful superconducting magnets. Such magnets are cooled by liquid helium to a temperature of about -269 degrees Celsius, at which point the copper cables suddenly lose all resistance to electricity. This means that far higher currents can be supported without the magnet burning up—and it also saves in utility bills. The Tevatron, which became operational in 1983, was soon colliding protons and antiprotons with a beam energy of over 1 TeV, or 1,000 GeV. The old bubble chambers had been replaced by huge detector halls, stacked with concentric cylindrical rings of sophisticated electronic detectors, each designed to measure the attributes of a different type of particle. Photographs of curving particles were replaced with computer graphics and reams of data that had to be analyzed statistically.

Even as the Tevatron was getting into its stride, physicists were setting their sights on a still larger machine—the Superconducting Super Collider, or SSC.

A Bigger Gun

The SSC had a planned ring circumference of 54.1 kilometers and an energy per beam of 20 TeV—almost three times greater than the LHC's maximum planned energy of 7 TeV. After a long campaign from physicists, the project was presented to Ronald Reagan in 1987 by Alvin Trivelpiece, who was the director of the Office of Energy Research. During his pitch, rather than focus on technical details, he put up a picture of a hay bale. Inside it, he said, were some billiard balls. "Without breaking the hay bale apart," he said, "how do you work out where the billiard balls are?"[29] Appealing to the President's love of firearms, he suggested that one way would be to use a BB gun to shoot into the bale—but the shots might not be strong enough to penetrate. "But what if you loaded up a rifle, stood back, and shot high-velocity bullets into the bale every half-inch or so?"

The SSC, it seemed to be implied, was like the biggest gun ever assembled. Of course, the bullets were on the small side—the total energy generated by the collisions would be about the same as striking a match—but no one dwelled on that. As the physicist Peter Higgs told science writer Ian Sample, "I think Reagan was thoroughly confused as to whether this thing was going to help him zap the Commies."[30]

The start of the SSC's construction in 1991, in a site south of Dallas, Texas, coincided with the end of the Cold War. At first, this seemed a good thing for the project, since the peace dividend meant that the U.S. government could employ defense companies such as Lockheed and General Dynamics to build magnet systems instead of bombs. The SSC even began recruiting Soviet scientists in order to prevent their nuclear expertise from being absorbed by other countries or sold to the highest bidder. This resulted in unlikely pairings of U.S. military contractors and ex-Soviet scientists. As one American told me, "One day we're getting ready to nuke them, the next we're working with them."

Although good for international harmony—one of the greatest achievements of accelerator physics is the way in which it has brought scientists from around the world to work together—this spirit of cooperation also removed some of the government's incentive for pursuing expensive physics projects. While the connection between the SSC and military applications was not explicit, high-energy physics had been associated with the development of new weaponry ever since the Trinity test. Because of this, as the science historian Daniel J. Kevles observes, "American physicists became a kind of secular establishment, with the power to influence policy and obtain state resources largely on faith and with an enviable degree of freedom from political control. What brought them to power is, to a considerable degree, what kept them there for most of the last half century—the identification of physics with national security."[31] Even if government leaders couldn't understand what the accelerators were actually for, apart from shooting the atomic equivalent of hay bales, they did understand that "seemingly impractical research in nuclear physics had led to the decidedly tangible result of the atomic bomb; thus, research in particle physics had to be pursued because it might produce a similarly practical surprise. In the context of the Cold War, particle physics provided an insurance policy that if something important to national security emerged unexpectedly, the United States would have the knowledge ahead of the Soviet Union."

In the American psyche—and the minds of its leaders—the race for supremacy in physics was therefore tied up with the battle for technological supremacy against its communist adversary. When the Soviet Union imploded, so did some of that spirit of discovery which had bankrolled the SSC to the tune of billions. As the Republican Senator Dave Durenberger put it, "If we were engaged in a scientific competition with a global superpower like the former Soviet Union, and if this project would lead to an enhancement of our national security, then I would be willing to continue funding the project. But . . . we face no such threat."[32]

The SSC had other problems apart from the expiry of the Cold War. Design changes had expanded the budget from an initial $6 billion over ten years to more than $8 billion (in 1990 dollars), and delays meant further increases. Politicians and the electorate were focusing on the ballooning national deficit, and the short budget and congressional cycles meant that the project was vulnerable to a bad year—unlike the LHC, which had secured long-term funding.

In 1993 the SSC budget was cut by Congress and construction came to an abrupt halt. Some $2 billion had already been spent and 20 percent of the tunnel excavated. A couple of thousand scientists, engineers, and supporting staff—including the author—found themselves without a job. Scientists in other competing areas were disappointed when the funds saved were absorbed elsewhere. The site is now unused, though the tunnel was featured in Jean-Claude Van Damme's 1999 movie, *Universal Soldier: The Return.*

In the United States at least, the collider race ended with a crash. The hopes and ambitions of the accelerator community transferred themselves to CERN's proposed Large (but not quite so large) Hadron Collider.

Chasing Unity

While the military connection helps to explain why high-energy physics had access to much greater funding than other types of science, it doesn't answer why scientists themselves would put such emphasis on building ever-bigger machines, or why the government and public in general would be willing to support projects for which they evinced such little understanding.[33] Commercial spin-offs from high-energy physics have been rare. It has been variously claimed that accelerator research has led to advances in everything from superconducting cables to tunnel construction to the World Wide Web (its HTML language was invented by Tim Berners-Lee while he was at CERN), but these were related to the technology used in laboratories rather than the actual physics. Solid-state physics, in comparison, has far less funding and prestige—the particle physicist Murray Gell-Mann called the area "squalid state physics"—but it has generated far more in the way of useful inventions and commercial applications, including computer chips.[34]

The mystique of high-energy physics no doubt helped—Einstein was aware of this effect, telling a Dutch newspaper in 1921 that "it is funny and also interesting to observe. I am sure that it is the mystery of non-understanding that appeals to them."[35] There is also the sheer love of advanced high-tech devices. However, the main reason for the support of accelerator projects is related to something even more powerful: the belief, which goes back to the Greeks, that knowledge is ranked in a linear way from fundamental truths to superficial appearances. If we can understand the atom, or whatever makes up the atom, then the rest will follow.

This belief in reductionism is in turn tied up with an aesthetic and almost spiritual desire for order and unity. In 1987, the physicist Steven Weinberg eloquently put the case for the SSC to Congress and to the public.

> There is reason to believe that in elementary particle physics we are learning something about the logical structure of the universe at a very, very deep level. The reason I say this is that as we have been going to higher and higher energies and as we have been studying structures that are smaller and smaller we have found that the laws, the physical principles, that describe what we learn become simpler and simpler. . . . There is simplicity, a beauty, that we are finding in the rules that govern matter that mirrors something that is built into the logical structure of the universe at a very deep level.[36]

Anthony Zee wrote that, "In our limitless hubris, we are beginning to feel that we are on the threshold of really knowing His thoughts."[37]

On a personal level I appreciate Weinberg's testimony, since it was instrumental in getting the SSC—which really was a fantastic project to be involved in—approved in the first place. However, I find it hard to understand what he meant by physical principles becoming simpler and simpler. To me, as one of the team working on the design of the superconducting magnets, the range of particles produced by the machines—and the mathematics required to model them—seemed bafflingly complex. Even Enrico Fermi told a student, "If I could remember the names of all these particles, I would have been a botanist."[38]

The only way to make that diversity go away, it seemed, was to continue the quest for more fundamental particles that would somehow make sense of it all. As we will see, this worked to an extent, but every time a new energy was attained, new particles or phenomena or questions seemed to pop into existence. The desire to build more powerful accelerators was like a kind of addiction—each hit would lead the scientific community closer to ecstasy, but the effect would soon wear off.

The Particle Explosion

In the 1940s, when accelerator science was emerging from its infancy, the number of known particles could be counted on two hands. There were the electron, proton, and neutron, which made up the atom; the massless photon, which made up light; plus a few exotic and unstable particles which had been discovered in cosmic ray experiments, but were soon also being generated by accelerators. With the advent of more powerful machines in the 1950s and 1960s, the list quickly expanded.

One of the first new particles to be discovered in an accelerator was the antiproton. The earlier discovery by Carl Anderson of the positron had suggested that the proton too could have an antimatter version. From Einstein's $E=mc^2$, it should be possible to produce a particle/antiparticle pair if the energy available exceeds the combined mass of the two particles. For example, when a high-energy photon (say from a cosmic ray) collides with a proton, the energy from the collision can be enough to produce an electron/positron pair, which is one way to produce antimatter. Another method is to just smash stuff together very hard.

Physicists computed that a beam of protons, when accelerated to an energy of 6 GeV and aimed at a stationary target, would supply the requisite energy to make an antiproton. The Bevatron at Berkeley—not by coincidence—was designed by Ernest Lawrence to have an energy of slightly more than that amount.[39] In a 1956 experiment, protons colliding with nuclei in a copper block were shown to produce antiprotons that could be detected by their momentum and velocity.[40] By the early 1960s, the antimatter equivalent of most particles had been discovered, which immediately doubled the number of known particles. Antiparticles, and even antiatoms, are now routinely produced at accelerator laboratories.

Next up was the neutrino, which Wolfgang Pauli had predicted was produced during radioactive beta decay, as discussed in Chapter 3. The fact that the proposed particle was extremely light, traveled at near the speed of light, interacted weakly with other forms of matter, and carried no charge meant that it could not be detected by conventional means. In 1956, Clyde L. Cowan and Frederick Reines used a nuclear reactor to produce enough neutrinos that their interactions with water molecules in a large tank could be detected. When a neutrino collided, very occasionally, with a proton in the water, it produced a neutron and positron. The positron then collided with an electron to annihilate in a burst of high-energy light radiation. We now know that neutrinos are the most ubiquitous of particles, with billions of them from nuclear reactions in the Sun alone flowing through your body at every second.

An interesting property of neutrinos is that they are all left-handed, in the sense that if you point the thumb of your left hand in the direction of the neutrino's travel, then your fingers will always curl in the direction of its spin. Other particles such as electrons can be either left- or right-handed. The so-called parity violation of neutrinos was a consequence of the discovery made in 1956 by the Chinese physicist Chien-Shiung Wu, or "Madame Wu" as she was often known, that the weak force is fundamentally asymmetric. Her finding was corroborated in a separate experiment led by Leon Lederman. This came as a severe shock to physicists, who had always assumed on the basis of aesthetics that the laws of physics would be perfectly symmetrical, so that a mirror image of the universe would be indistinguishable from the real one. But a mirror image of a left-handed neutrino is a right-handed neutrino, and that appears not to exist.[41] As Wolfgang Pauli exclaimed, "I cannot believe God is a weak left hander!"[42]

In 1962, Lederman, Melvin Schwartz, and Jack Steinberger showed that neutrinos also come in different types. The experiment, carried out at Brookhaven's new AGS machine, used a proton beam to collide with a target and generate a shower of particles which included pions. The pions were filtered out magnetically and directed 70 feet towards a steel wall, which weighed 5,000 tons and was constructed from old battleship plates. Pions are highly unstable, so they decayed on the way into muons (a heavy version of an electron) and neutrinos. The muons were screened out by the wall, but the muon-neutrinos, as they became known, passed right through and were picked up by a specially designed detector.[43]

As accelerators continued to grow in power through the 1960s and 1970s, the number of particles increased in step. In an attempt to impose some order, physicists grouped them into separate families based on properties such as mass, charge, spin, and so on. Leptons, named after the Greek word for fine or small, included the electron, the muon, and the heavier tau particle, together with their corresponding neutrinos. Hadrons, from the Greek word for large or massive, were divided into baryons and mesons. Baryons included the proton and neutron. Mesons included pions, kaons, and other particles with intermediate mass. These in turn could be broken down into finer categories. Pions, for example, which took their name from the Greek letter pi, came in the forms π^0, π^+, and π^-, depending on their charge. Many particles were so incredibly short-lived that they could not be detected directly, but only inferred from the way they decayed and interacted with other particles. Since quantum collisions are inherently random, scientists had to statistically analyze large numbers of events in order to determine the underlying processes.

Kaons had a special property, called strangeness by Murray Gell-Mann, which prevented them from decaying as quickly as expected.[44] Their lifespan was about one ten-billionth of a second—which, if they were moving at near the speed of light, allowed them to traverse some 3 centimeters before expiring. The kaons were also only produced in pairs. Gell-Mann therefore proposed that strangeness was a quantum number, like charge or spin, that was subject to a conservation principle during interactions other than those involving the weak force. For example, if a nuclear collision produces a new particle with strangeness +1, it must normally balance that with a new particle of strangeness -1. Such quantum numbers soon proliferated to include baryon number, hyperon number, and so on. The total number of hadrons that had been discovered and predicted soon exceeded 200, and included names such as the charmed double bottom Omega.

The Pythagorean Way

The supply of Greek letters used to symbolize the particles was threatening to run out and something had to be done. The first steps in finding a periodic table for subatomic particles were taken by Gell-Mann and Yuval Ne'eman in 1962. They independently arrived at a classification scheme which Gell-Mann named the Eightfold Way, after the Buddha's eightfold path to truth. It was based on the observation that, when arranged spatially according to strangeness and charge, the hadrons appeared to fit naturally into certain patterns of eight and ten particles, as shown in Figure 6.2.

The flippant allusion to the Buddha is misleading; the model for these patterns was surely Pythagoras, who lived around the same time. The Pythagoreans visualized the properties of numbers using exactly these types of spatial patterns. The baryon decuplet, shown in the panel on the right, is an inverted tetractys—a figure which the Pythagoreans considered to be sacred. Indeed, an obituary of Yuval Ne'eman in the CERN newsletter noted that "Yuval saw himself as a disciple of the Pythagorean tradition in theoretical physics . . . that physical phenomena can be described by simple and beautiful mathematics. Yuval's own work is a prime example of the strength of this principle."[45] These patterns therefore represent the Pythagorean, not the Buddhist, way to truth (these being quite different things).

Just as the discovery of the rhythmical octave structure in the chemical elements suggested some type of underlying structure for the atom, so the discovery of these geometrical patterns suggested a structure for the subatomic particles. In 1964, Gell-Mann and George Zweig found

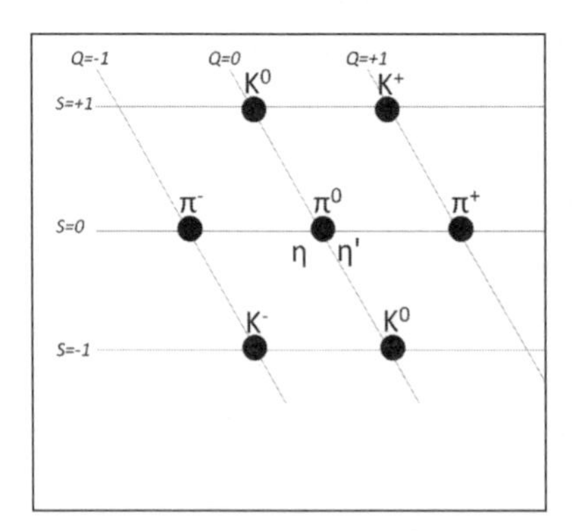

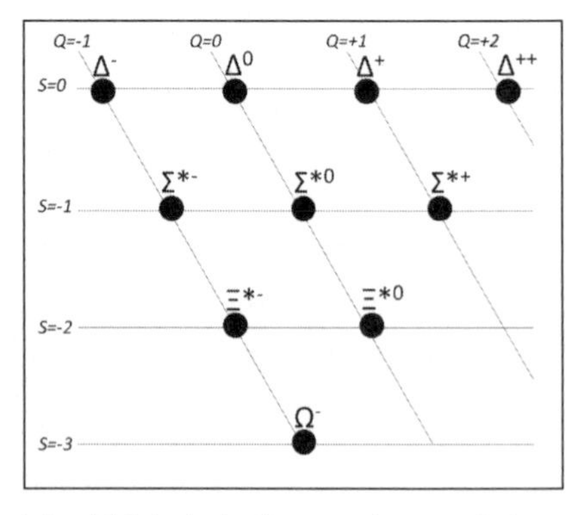

Figure 6.2 Left panel shows the meson octet, which includes kaons, pions, and eta mesons. Particles along the same horizontal line share the same strangeness quantum number, denoted S. Particles on the same diagonals share the same charge, Q. Baryons with spin 1/2 fall into a similar pattern. Right panel shows the baryons with spin 3/2. Source: Author

that the patterns could be explained if the hadrons were viewed, in an abstract sense, as being made up of entities which Gell-Mann dubbed quarks and Zweig named aces.[46] Gell-Mann's term, which came from the line, "Three quarks for Muster Mark!" in James Joyce's *Finnegan's Wake*, won out. (His talent for finding catchy names extended even to himself—he added the hyphen to his surname.)

Quarks came in three types, or flavors: up, down, and strange. Baryons were made up of three quarks, so for example a proton was two ups and a down, while a neutron was two downs and an up. Mesons, which had smaller masses, were made up of a quark and an antiquark. A positive pion was an up quark plus a down antiquark, while a negative kaon was a strange quark plus an up antiquark.

While the new scheme certainly helped to make sense of the particle zoo, it didn't do a perfect job of explaining the observed masses of mesons. Within a year, Sheldon Glashow and James Bjorken had proposed a fourth flavor of quark. They called it charm, as Glashow explained, because "we were fascinated and pleased by the symmetry it brought to the subnuclear world."[47] It was later joined in the 1970s by two more quarks with very heavy masses. These were known alternately as truth and beauty or top and bottom. The latter designation eventually caught on.

Quarks could transform into other quarks in a process mediated by the weak force. In beta decay, a neutron (two down quarks and an up quark) could transform into a proton (one down quark and two up quarks) plus an electron and an electron antineutrino. So the weak force could change a down quark into an up quark,[48] while a charm quark could be transformed into a strange quark (especially after having too much to drink at parties).

Gaps in Mendeleev's periodic table led him to predict the existence of new elements. In the same way, there were some combinations of quarks that had not yet been observed. One such gap was the Omega minus particle—made up of three strange quarks—which is to be found at the bottom of the decuplet in Figure 6.2 (i.e., the peak of the tetractys). Its discovery in a bubble chamber at Brookhaven in 1964 was strong evidence that the theorists were on the right track.

Truth and Beauty

At first, quarks were viewed as purely mathematical constructs; as with atoms in the nineteenth century, there was little direct evidence for their existence. In 1968, some empirical evidence appeared courtesy of the Stanford Linear Accelerator Center, whose electron beam was the perfect instrument for probing the structure of protons. Just as Rutherford had used scattering to show that the atom was made up of a charged nucleus surrounded by electrons, so the team at SLAC found that when they fired electrons at protons in a target and observed the scattering statistics, the protons appeared to be made up of smaller objects. (The experiment was similar to feeling a bag of marbles to guess how many it contains.) These were later identified with up and down quarks. However, the quark theory was only fully accepted by the majority of physicists in 1974, with the experimental discovery at both SLAC and Brookhaven of the charm quark. The final quark to be discovered was the top quark, which was found at Fermilab in 1995 and turned out to have a mass almost equal to that of a gold atom.

The reason why quarks took so long to be accepted by the physics community is that they were never observed on their own in the wild and only appeared bound together in groups of two or three. The force which held the nucleus together had long been a puzzle to theorists. The protons in the nucleus had the same positive charge, and so they would repel one another. It was also known from nuclear experiments that the force which bound them to the nucleus was very powerful.

According to quantum field theory (see Box 6.2), forces are transmitted by the exchange of virtual particles. In the case of electromagnetism, for example, charged particles such as electrons are surrounded by a cloud of such virtual photons, which are constantly flashing in and out of existence. Theorists also devised a quantum field theory for the weak nuclear force: here, the force-mediating particles were not weightless photons, but massive particles denoted W^+, W^-, and Z^0. The fact that these particles had mass meant that they could only exist for a short period before reaching the time limit set by the uncertainty principle. They therefore acted over only a tiny range, less than the size of a proton.

Reasoning along the same lines, theorists in the 1970s proposed that the strong force was transmitted between quarks by particles—dubbed gluons—with a range of about the size of a proton. The sign of the force depended on yet another quantum number, called color by Gell-Mann, for quarks and gluons. This was analogous to charge but could take on three values, arbitrarily denoted red, green, and blue. Just as atoms were charge neutral, hadrons were color neutral, in the sense that the colors of the three quarks were complementary and therefore added to white. Protons and neutrons were held together in the nucleus by the attraction between the composite quarks.

Unlike the electromagnetic or weak force, the strong force was more like a rubber band. Quarks could move freely so long as they were confined within its range; but if they were broken apart, the energy needed to snap their bonds was so high that new quarks and antiquarks would be created. These in turn would immediately bind with the free quarks to form hadrons.[49] Thus the result of a high-energy collision was not separate quarks, but a jet of new hadrons. Many of the accelerator experiments in the 1970s and 1980s were directed at finding the characteristic jets generated by quarks and gluons.

Box 6.2

Perfect World

For the last twenty years of his life, the French philosopher Voltaire lived in a village called Ferney, which today lies above the LHC ring. In his most famous work, *Candide*, the protagonist Candide is mentored by Dr. Pangloss, an expert in "metaphysico-theologo-cosmolo-nigology" who teaches that we live in "the best of all possible worlds."[50] The book was a satire on the position of the philosopher and co-inventor of calculus Gottfried Leibnitz, who saw the world as the solution to a kind of optimization problem, with God as the master mathematician.

The idea of our world being chosen, subject to some criterion, from a variety of equally plausible alternatives has long been influential in physics. For example, the refraction of light as it passes through a substance such as glass can be modeled by assuming that light chooses the path that minimizes the total transit time. If you want to know the path taken by a light ray to get from point A to point B, you just search over the possible alternatives to find the fastest, taking into account that the speed depends on the medium and is fastest in a vacuum. Light chooses the path that is the most economical and perfect.

Similarly, the path of a moving object under a force such as gravity can be found by computing a quantity known as the Lagrangian. This is the difference between the kinetic energy associated with its motion and the potential energy that it has by virtue of its position. If this quantity is summed (integrated) at all points along the trajectory, giving a number known as the action, then the path taken is the one which gives the minimum action. In some sense, it is the one which most efficiently transforms force into movement.

In the 1930s, Paul Dirac worked out a version of the Lagrangian that accounted for quantum and relativistic effects. And after the war, the American physicist Richard Feynman used the method to produce a detailed quantum field theory for the electromagnetic force.

Feynman was present for the Trinity test and claimed that he was the one person to have viewed it without wearing protective glasses, relying instead on the windows of a truck and the laws of refraction to screen out ultraviolet radiation. He had a similarly fearless attitude towards physics, using his intuition to come up with new ways of visualizing quantum interactions which the old guard found offensive. His version of quantum field theory described the force between electrons as transmitted by virtual photons whose paths had an associated Lagrangian.[51] Because this was a quantum, probabilistic system, the electrons and photons were not limited to one path, but could potentially explore all possible paths.

For example, an electron could emit a virtual photon and then reabsorb it; or (less likely) the emitted photon might produce a new electron/positron pair, which would then join and annihilate within about 10^{-21} seconds to produce a photon, which would again be reabsorbed by the electron; and so on. The probability of a final state being achieved would depend on a weighted sum of all the actions, which takes into account their probability.

In the quantum world, anything can happen. But when averaged over a large number of events, Dr. Pangloss's statement that we live in an optimal world would still be correct—at least in physics.

The Standard Model

The inventory of particles and forces in the Standard Model, as Steven Weinberg christened it in 1971, therefore reads as follows. There is the electron and its neutrino, plus two more "generations" of leptons: the muon and muon-neutrino, and the tau and tau-neutrino. The hadrons are composed of three generations of quarks: up/down, charm/strangeness, and top/bottom. Each of these matter particles has an antimatter version, for a total of twenty-four. In addition, there are the force-carrying particles: the photon, the three weak force bosons, and eight types of gluon. So in all there are thirty-six particles. If the different charges of quarks are taken into account, then the total expands to sixty particles.

As discussed further in the next chapter, a major development in the Standard Model was the unification of the electromagnetic and weak forces within a consistent mathematical framework by Sheldon Glashow, Steven Weinberg, and Abdus Salam in 1967. This electroweak theory relied on the introduction of a new particle, known as the Higgs boson, which is responsible for giving particles their mass. One of the main goals of the LHC (as with the SSC before it) was to detect the Higgs boson (as discussed below, this was apparently achieved in July 2012). Finally there is the graviton, which transmits the gravitational force but is beyond the range of any accelerator. Including these two hypothesized particles, the grand total for the Standard Model is sixty-two. For comparison, that is one less than the number of elements that Mendeleev knew of in the 1860s when he invented the periodic table.

Mathematical symmetries—such as shared masses or other attributes—mean that the actual number of parameters that have to be adjusted by hand in the Standard Model is somewhat smaller, totaling about twenty (depending on how you count). These include the mass and charge of the electron, the mass of the quarks, and so on. But from an aesthetic point of view, this number is still obviously not very satisfactory—and it's a long way from unity. Albert Einstein wrote that "I cannot imagine a unified and reasonable theory which explicitly contains a number which the whim of the Creator might just as well have chosen differently, whereby a qualitatively different lawfulness of the world would have resulted. . . . A theory which in its fundamental equations explicitly contains a constant [of Nature] would have to be somehow constructed from bits and pieces which are logically independent of each other; but I am confident that this world is not such that so ugly a construction is needed for its theoretical comprehension."[52] The Standard

Model fell well short of this standard of perfection. As Leon Lederman put it, "The drive for simplicity leads us to be very sarcastic about having to specify twenty parameters. It's not the way any self-respecting God would organize a machine to create universes. One parameter—or two, maybe. An alternative way of saying this is that our experience with the natural world leads us to expect a more elegant organization. . . . The problem is the aesthetics."[53] Or, in the words of Chris Llewellyn Smith from CERN, "While the Standard Model is economical in concepts, their realization in practice is baroque, and the model contains many arbitrary and ugly features."[54]

The idea that physics is making progress towards ever greater unity and simplicity is therefore a cliché which is not entirely borne out by the facts. As physicist Marcelo Gleiser observes, "there is indeed some delusion going on. The hints of a final theory . . . are not as obvious or even as circumstantial as it is widely believed."[55] Solid-state physicist Robert Laughlin writes of the reductionist paradigm that "mythologies are immensely powerful things, and sometimes we humans go to enormous lengths to see the world as we think it should be, even when the evidence says we are mistaken."[56] We are imposing our ideas about beauty on reality. Theorists have certainly been successful in spotting patterns in nature at the subatomic level. But rather than being proof of some deeper Pythagorean structure or symmetry, it may be that such patterns are better described as emergent phenomena which appear from time to time in all levels of complex systems.

The definition of emergence is somewhat hazy, but the basic idea is that a system can exhibit properties which cannot be understood by reducing the system to its components. The term first emerged in the early twentieth century when it was used to describe living systems by philosophers such as Samuel Alexander, who wrote in 1920, "Physical and chemical processes of a certain complexity have the quality of life. The new quality life emerges with this constellation of such processes, and therefore life is at once a physico-chemical complex and is not merely physical and chemical. . . . The existence of emergent qualities thus described is something to be noted, as some would say, under the compulsion of brute empirical fact. . . . It admits no explanation."[57]

Emergent properties are in fact characteristic of complex systems in general, including biological, social, or physical systems. As will be discussed further in Chapter 8, they are incomputable in the sense that they cannot—even in principle—be determined from a knowledge of the components alone. Sometimes emergent properties are so exact that they appear to represent a fundamental law, but this is misleading. Phase changes, such as the freezing of water to ice, are exact emergent properties that cannot be deduced from a study of individual molecules.[58] Even the concept of an object's solidity melts into nothing when the number of atoms is too small. Another phase change exploited by engineers is that of superconductivity, which occurs when some materials are cooled below a critical temperature. As Laughlin notes, superconductivity is a "tendency of electrons to lock arms and move as one gigantic body" that breaks down when there aren't enough electrons.[59] In other cases, as will be discussed later for things like clouds, the emergent properties are inexact and unpredictable.

Another example of an exact emergent property is the frequency and speed of sound waves passing through a metal bar. As with quantum wave/particle duality, the waves are equivalent to an "emergent particle" known as a phonon, which has a well-defined momentum. If the sound waves are weak enough, the sound becomes quantized into discrete pulses just as light does. Of course, these particles might not seem to exist in the same hard, physical sense that the bar does—but then neither do atoms, as shown by quantum physics. The diversity of particles and intriguing patterns that appear in the Standard Model could similarly be the emergent result of some finer-scale processes.

This doesn't make those particles or patterns any less real, but it does change the way we look at them. As we will discuss further in Chapter 10, emergent properties pose a challenge to both the reductionist program and the mainstream scientific aesthetic, because they imply that what counts is less atoms or particles than the collective organizational principles which unexpectedly emerge when enough of these particles are brought together. We can't derive superconductivity from quantum physics, but we can use it to build magnets; we can't compute a snowflake from a model of its atoms, but we can catch one on our tongue. The beauty is in the emergent phenomenon, which disappears when we try to anatomize it.

The evident lack of unity displayed by the smallest known components of the universe has certainly not dissuaded theoretical physicists from the goal of a perfectly unified theory. Many continue to believe that the universe really is perfectly symmetric and ordered—or at least that it used to be. In the next chapter, we will show how aesthetic criteria such as harmony, unity, and symmetry changed from being aids in the comprehension of nature to prescriptive rules governing a pristine, literally weightless world whose beauty can only be imagined.

7

Broken Mirrors

Beauty is bound up with symmetry. . . . Symmetry, as wide or as narrow as you may define its meaning, is one idea by which man through the ages has tried to comprehend and create order, beauty, and perfection.

Hermann Weyl, *Symmetry*

Fundamental physicists are sustained by the faith that the ultimate design is suffused with symmetries. . . . They hear symmetries whispered in their ears.

Anthony Zee, *Fearful Symmetry*

I reiterate again that the primary lesson of physics in this century is that the secret of nature is symmetry.

David J. Gross, *Gauge Theory—Past, Present, and Future?*

Scientists have always associated symmetry with both beauty and truth. The Greeks believed that the stars and planets moved in perfect circles around the Earth because circles were purely symmetric and therefore perfect. Today, scientists have taken symmetry to new lengths—even proposing that each known particle has a supersymmetric (but so far undetected) partner. But many things in nature—including for example the weak nuclear force, the mix of matter and antimatter in the universe, or our own DNA—are asymmetric. This chapter argues that it is time for scientists to break the mirror. For as Sir Francis Bacon noted, "There is no excellent beauty that hath not some strangeness in the proportion."[1]

One of the main aesthetic principles in science and elsewhere is the concept of symmetry. Experiments have shown—and plastic surgeons can also attest—that we largely assess the attractiveness of faces on the basis of left/right symmetry. One study even claimed that people with highly symmetric features are more selfish and individualistic, presumably because they can get away with it (under the surface, of course, the layout of even their organs is less tidy).[2] However, facial symmetry should not be completely perfect; we find that slightly creepy, perhaps because it reminds us of a machine.[3] As a character in Thomas Mann's *The Magic Mountain* reflects while gazing on a snow crystal, "Life shuddered at such perfect precision."[4] Marilyn Monroe and later Cindy Crawford were both considered by many to be "perfect models" of the glamour variety, but were also famed for their symmetry-breaking facial moles, which in their case were called beauty marks.[5]

Symmetry of body features is important for mate selection in many other species as well, which is perhaps unsurprising given that major deviations from symmetry can be a sign of injury, disease, or genetic impairment. Symmetric patterns also appear widely in most cultures as symbols or architectural devices; however, they seem to play a particularly strong role in Western culture, where they have often been viewed as the bedrock of reality.

One of the earlier versions of a Theory of Everything was that presented by Plato in his dialogue *Timaeus*. He describes how the universe was created by a Demiurge—a kind of divine cosmic architect—using the perfect Forms as a template. The Earth and the universe itself were made spherical, on the grounds that "there is no shape more perfect and none more similar to itself"—in other words, the sphere looks the same from every angle and can be rotated by any amount without changing its appearance. All of the components of the universe were arranged according to numerical harmony. The four elements were made of particles having the shapes of the Platonic solids, which are highly symmetric and can each be inscribed in a sphere. Although Plato didn't directly mention symmetry—the word as we know it did not come into use until the eighteenth century—it's clear that he had it in mind.

Plato was influenced by Parmenides, who as discussed in Chapter 4 had similarly described the What Is as "perfected from every side, like the bulk of a well-rounded globe, from the middle equal every way."[6] He in turn was influenced by the Pythagoreans, who saw the sphere as the most symmetric and therefore most perfect shape.

Spheres and other symmetrical objects have continued to be kicked around by philosophers ever since, like a kind of mental football. The Greek models of the cosmos were all based on circles or spheres, on the grounds that, as Ptolemy put it, they alone are "strangers to disparities and disorders."[7] In the thirteenth century, the most popular book on astronomy, written by John of Holywood and called simply *The Sphere*, argued that the universe was in the shape of a giant globe. Kepler's *Mysterium Cosmographicum*, which set itself the aim of deducing "the shape of the Universe and the unchangeable symmetry of its parts," again saw the design of the universe as being based on spheres and the Platonic solids. Einstein initially assumed that the universe was a static sphere and introduced his cosmological constant to keep it that way.

Symmetry is so attractive that we sometimes consciously or unconsciously screen out imperfections. The pioneer of Victorian high-speed photography, Arthur Worthington, made a number of studies of splashing water (Figure 7.1). As he wrote, "I have to confess that in looking over my original drawings I find records of many irregular or unsymmetrical figures, yet in compiling the history it has been inevitable that these should be rejected, if only because identical irregularities never recur. Thus the mind of the observer is filled with an ideal splash—an 'Auto-Splash'—whose perfection may never be actually realized."[8] Part of the appeal is that symmetrical shapes are easy to analyze mathematically. An old joke relates how a farmer brought in some scientists to find why milk production in his farm was so low. After much computation, the lead scientist announces that he has the solution, but it only works in the case of spherical cows in a vacuum.

Today, of course, science has become incredibly sophisticated. We now think that everything in the universe is based on mathematical laws, which are described by the Standard Model, which many believe is a special low-energy case of a more sophisticated supersymmetric string theory. Interestingly, though, the spheres haven't gone away—in fact, they seem to have multiplied. Physical laws are like oranges at the supermarket: everyone wants one that looks perfectly round. So are these spheres real, or—like Aristotle's crystalline spheres—are they a projection of our desire for order and perfection?

The Circle Force

On first glance, it might seem that something like quantum electrodynamics is unrelated to circles or spheres. But a property of its equations is that they incorporate a symmetry which is identical to that of a circle. This in turn is related to a property of waves.

In quantum theory, particles are associated with wave functions. The simplest type of wave is a sine wave, of the sort shown in Figure 7.2. The amplitude is the wave's maximum height and the phase is a number which describes its location in space. Quantum wave functions are more complicated because their value at any point is a so-called complex number, which has a "real" component and an "imaginary" component. (The latter is related to the square root of a negative number, which is why it is called imaginary—normally if you multiply a number by itself you get a positive number. The square root of negative one is denoted by the letter *i*. The amplitude of a complex number is determined by applying the Pythagorean theorem to the two components: sum the squares of each and take the square root. Note the use of the

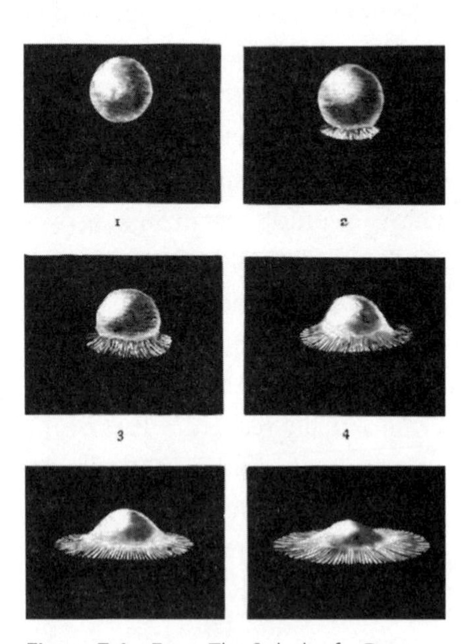

Figure 7.1 From *The Splash of a Drop*, by Arthur Worthington. Compare with the Trinity nuclear blast, shown in Figure 5.2. Source: Reprinted from A.M. Worthington, *The splash of a drop* (London, Society for Promoting Christian Knowledge, 1895).

word "complex" here is not related to complex systems.) One can picture the imaginary component as being in the direction perpendicular to the plane of the figure.

The phase has the property that if you increase it by 360 degrees, the wave ends up where it started—just as the hands on a clock face repeat their positions every twelve hours. The phase is often called a phase angle, and it is visualized as an angle on a circle. It turns out that almost any type of wave—including highly localized waves, which correspond to particles that are localized in space—can be represented as the sum of a series of similar sine waves.

Now, in quantum electrodynamics, the equations have the property that if the wave functions of a system under analysis are all locally shifted in phase by exactly the same amount, then the underlying physics remain the same. For example, if two light waves interact, as in a double-slit experiment, then what counts is only their phases relative to one another, and this will be unchanged. The mathematician Hermann Weyl in 1918 called this a gauge symmetry.[9]

The situation is analogous to a spatial symmetry, where for example the results of an experiment should not be affected by a shift in location. As was discussed in Chapter 3, Emmy Noether showed that global symmetries implied conservation principles. Here the symmetry implies

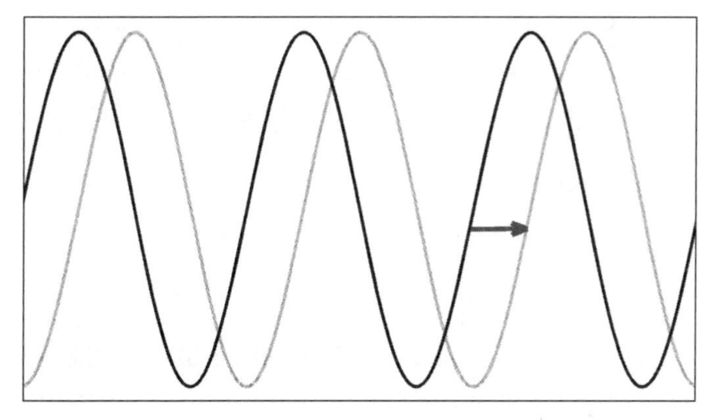

Figure 7.2 Plot of a sine wave. A change in phase, indicated by the arrow, corresponds to a horizontal shift in position. If the phase is increased by 360 degrees, the wave overlaps with itself and so remains unchanged. Source: Author

charge conservation—the idea that electrical charges can be neither created nor destroyed. But in this case the symmetry is local, in the sense that it can be allowed to vary smoothly and continuously over spacetime (in the same way that Einstein's general relativity allows for local changes in gravity). This is a much stricter condition, and Weyl showed that if we write out the equations for a generic field with this property of local phase invariance, then it turns out to be the same as the equations for electromagnetism. The force is in a sense defined by its symmetries; the only thing to be left unspecified is its strength.[10]

In mathematical terms, a symmetry is the same as invariance under a particular operation (see Box 7.1). A square is invariant under rotations of 90 degrees around its center, but a circle is invariant under any rotation. We can therefore say that the electromagnetic equations have the same symmetries as a circle (physicists usually work in complex numbers, so treat this as a circle in the complex plane with one real and one imaginary axis—but it is equivalent to an ordinary circle in two dimensions). The mathematical set of symmetry transformations, namely phase angles, are known as U(1), for unitary group in one variable—or more simply as the circle group.

We can therefore say that, viewed through symmetry goggles, electromagnetism is basically a circle. Of course, the circle description of the electromagnetic field does not tell us anything new about electricity, magnets, or light, and certainly isn't a very useful way of describing the practical effects of electromagnetism. It was therefore initially treated as a mathematically elegant contrivance. That would soon change.

The Sphere Force

Because symmetry is the same as invariance under certain operations, a symmetry in a field is the same as blindness to changes of a particular sort. A circle is blind to rotation. The electromagnetic force is blind to phase change. Similarly, the weak nuclear force is blind to the difference between electrons and neutrinos, in the sense that the equations for the weak force (specifically the Lagrangian) are invariant to transformations that swap them around. Electrons and neutrinos of course have very different masses, but mass is treated separately, as we will discuss below.

Box 7.1

Squaring the Universe

The Pythagoreans reasoned about numbers in terms of arrangements of pebbles. Square numbers were those such as four, nine, and sixteen, which can be represented by a square array. These square numbers, which were believed to have special properties of solidity and integrity, were associated with the male odd numbers, because they can be expressed as the sum of consecutive odd numbers: $4 = 1 + 3$, $9 = 1 + 3 + 5$, $16 = 1 + 3 + 5 + 7$, and so on. The proof is geometric; a square of a given size can be extended by adding an L-shaped row, and the number of pebbles required at each step is the next odd number.

As Aristotle pointed out in *Categories*, squares differ from oblongs because the ratio of a square's two sides always remains fixed at one. For example, the oblong number two can be represented by one row of two, with a ratio of 1:2. When a layer of pebbles is added to the top and side, there are two rows of three. The ratio therefore changes to 2:3, and so on at each step. Squares are therefore fixed and limited while oblongs are unstable and plural.

The symmetry of the square means that it can be described by just a single number—the length of one side—in the same way that a circle can be specified by its diameter. Symmetry can therefore be viewed as a tool to reduce plurality to unity. The same trick is used by physicists in order to reduce complex phenomena to simple and elegant equations.

Mathematicians view symmetry in terms of invariance under a particular transformation, such as rotation or reflection—in other words, they see it in terms of mathematical operations on objects rather than the objects themselves. If a square is rotated by 90 degrees around its center, or by any other multiple of 90 degrees such as 180 degrees or 270 degrees, then it does not change. The most symmetric shape is a circle, which is invariant or stable under any rotation.

Symmetry is therefore associated mathematically and aesthetically with the ideas of stability, unity, and the limited, while asymmetry is associated with instability, plurality, and the unlimited. These associations appear in the language used by scientists. For example, the word "normal" is from the Latin *norma*, meaning square. Physicists use the term renormalization to refer to the removal of infinities from equations, making them limited. As will be discussed in the next chapter, economists use the statistical technique known as the normal distribution (also called the bell curve) to reduce complex variability to a single number.

Symmetry is a way to make things normal and square. Which raises the question, of course, of whether the universe itself is normal and square.

In 1961, Sheldon Glashow developed a theory of the weak force, forming the basis of its unification with electromagnetism, in which its symmetries were denoted by the special unitary group in two variables, or SU(2).[11] This group is mathematically related to the rotations of a sphere, which are more complicated than the rotations of a circle. If you rotate a circle by an angle, say 30 degrees, and then by another angle, say 60 degrees, then the total rotation of 90 degrees would be the same as if you reversed the order (rotating first by 60 and then by 30 degrees). With spherical rotations, that isn't the case—if you reverse the order of two rotations that are not in exactly the same direction or at right angles to each other, then you end up with a different state. (Note that the rotations here are purely mathematical; they have nothing to do with actual movement in space.)[12]

These symmetries again place restrictions on the equations of the weak force. And in the same way that the U(1) symmetry specifies most properties of electromagnetism, when applied to quantum field theory the SU(2) symmetry turns out to specify most of the properties of the weak force. Instead of being mediated by a photon, it is mediated by three particles of spin 1, which unlike photons also interact with one another.[13] The strength of the force is not specified by the symmetry condition, but the measurement of a single weak interaction is enough to fill in the blanks—in the same way that knowing the size of one side of a square allows you to complete the rest.

While it seemed natural for physicists to model the weak force in terms of its symmetries, there was only one problem: according to theory, the three force-mediating particles shouldn't have any mass. In quantum field theory, particles interact by exchanging virtual particles, rather like two people tossing a ball back and forth. The range of the force depends on the mass of the object being tossed: the lighter it is, the farther it can go. Because photons have zero mass, the electromagnetic force has infinite range. If the particles that mediated the weak force also had zero mass, then the weak force would also have infinite extent, while it was known instead to be limited to a distance less than the size of a proton. This implied that the force-mediating particles were very heavy.

In fact, there was nothing in the math to say that *any* particles should have mass, or even exist except as potential excitations in a quantum field. The universe it described was as weightless, ethereal, and static as Plato's Forms.[14] So where did particles obtain their different masses?

One explanation, known as the Higgs mechanism, was suggested by Peter Higgs and others in the early 1960s.[15] This proposed that a new field and its associated particle pervades space and interacts with other particles in such a way that it grants them mass. The new particle became known as the Higgs boson—or, as Leon Lederman called it in a book, the God particle.[16]

Broken Symmetry

In 1993, the UK science minister hosted a competition to explain to the general public how exactly the Higgs mechanism worked. One of the winners was David Miller, a physicist from University College London. His original metaphor involved a cocktail party full of men. When an attractive woman enters the room, the men form a cluster around her, which slows her progress. In the same way, space is permeated by the Higgs field. As a particle passes through, it causes an excitation in the space—manifested by the Higgs boson—which slows its progress, giving it the effect of having mass. (In his official entry the cocktail party was replaced by a political function and the woman became Margaret Thatcher.[17])

Because different particles have very different masses, it is clear that the Higgs field interacts in a highly asymmetric manner. Why are muons heavy, electrons light, and photons massless? Here we have to distinguish again between the symmetry of the equations and the symmetry (or lack thereof) of the particular solutions. In the real world, symmetries get broken.

A common analogy is a pencil resting on its tip. The equations governing the motion of the pencil are symmetric, in the sense that if it is slightly perturbed it could fall in any direction. However, the particular solution is asymmetric: it falls in one direction only, thus favoring one direction over all the others, even though that asymmetry appears nowhere in the equations. This "symmetry breaking" introduces a random component into the calculation, because it can happen in many different ways. Thus the Higgs field could have been symmetric once upon a time, but at some point during the birth of the universe it collapsed by chance to a particular configuration.

A perhaps more informative interpretation of the Higgs process is to see it as a phase transformation, similar to the freezing of water. When water is in the liquid state, it is highly symmetric, in the sense that water molecules are distributed uniformly with no preferred direction. In ice crystals, however, water molecules are arranged in a neat hexagonal lattice. They are still very symmetric (which is why snowflakes also have a hexagonal symmetry), but they are less symmetric than water (which is why snowflakes are all different). But in this terminology, which is also the norm in solid-state physics, the emphasis is less on symmetry than on the shift in emergent behavior as molecules organize themselves into a new configuration with different properties. When water freezes, it becomes less dense. When the Higgs particles froze, the universe suddenly put on weight.

Of course, asymmetry does not necessarily imply a broken symmetry—it might just be asymmetric. However, the discovery in particle accelerators in the 1970s and 1980s of the W^+, W^-, and Z^0 weak-force-carrying particles, as predicted by the theory, was an impressive vindication of the SU(2) approach (their masses could be estimated by their range). While the description of electromagnetism in terms of U(1) didn't tell physicists anything new about nature, the same technique applied to the weak force had made an accurate prediction. The Higgs field only gained indirect support (Glashow compared it to a toilet; you might not be proud of it but something like it was necessary), but the idea soon pointed the way towards a greater kind of unification and symmetry.[18]

Color Blind

As discussed in the previous chapter, the goal of science has traditionally been to unify disparate phenomena, and the Higgs boson offered a way to do this. In the mid-1960s, Steven Weinberg and Abdus Salam independently showed that by using the boson to break Glashow's symmetries, it was possible to construct a model in which the weak and electromagnetic forces are two aspects of a single unified force, known as the electroweak force.[19] While these two forces have little in common today, they could have been identical at a time when energies were sufficiently high—for example, shortly after the big bang. Steven Weinberg wrote, "This is just the sort of thing physicists love—to see several things that appear different as various aspects of one underlying phenomenon. Unifying the weak and electromagnetic forces might not have applications in medicine or technology, but if successful, it would be one more step in a centuries-old process of showing that nature is governed by simple, rational laws."[20]

The success of electroweak theory meant that, when quarks were discovered in the 1970s, it was used as a template for a model of the strong force. According to quantum field theory, hadrons such as a proton are made up of quarks, which come in their three colors. The symmetries of the strong force are the operations which permute the locations of the three colors. If the weak force is blind to the difference between electrons and neutrinos, then the strong force is blind to colors. This set of mathematical transformations is referred to as SU(3) (here the comparison with higher-dimensional spheres becomes somewhat tortured).[21] The symmetries of this group are what generate the octuplet and decuplet patterns in Figure 6.2. The force is mediated by eight spin-1 particles, which are the gluons.

A number of physicists, including David Gross and Frank Wilczek, noticed that the strong force had a property known as asymptotic freedom.[22] Unlike the other forces, the strong force became weaker rather than stronger at very small distances. This allowed quarks to move freely so long as they were confined within a hadron. It followed from this that the force also became weaker at very high energies or temperatures. Extrapolating their calculations out to very high energies, the strength of the strong and electroweak forces came together at almost—but not quite—the same point. It was postulated that the forces were united at the time of the big bang and only emerged as separate entities when the universe cooled. First the strong force and the electroweak force would have split apart and then the electroweak would have divided into its electromagnetic and weak components.

Supersphere

The Standard Model was simply the combination of the electroweak and strong theories. It described key properties of the known particles and forces in terms of the three separate symmetry groups, U(1), SU(2), and SU(3), with only parameters such as the masses of particles and the strengths of the forces remaining to be adjusted by hand. Because it was a quantum field theory, the model had to approximate the result of interactions by summing over the quantum possibilities, and there were questions about mathematical consistency, but the answers were extremely accurate.

While the Standard Model seemed fairly elegant and sphere-like, at least in terms of its underlying symmetries, it left many questions unanswered. Why were there three separate symmetry groups, glued together in a rather awkward way, instead of just one? How was it that the strengths of the fields and masses of particles were set in such a way that everything worked? It wasn't long before physicists wondered if the three symmetry groups could themselves be unified into a single larger group—a supersphere. Perhaps the Standard Model was a manifestation at relatively low energies of a deeper, even more beautiful theory.

Unification was like a drug and physicists were hooked. In 1973, Sheldon Glashow and Howard Georgi proposed the first Grand Unified Theory, or GUT. This modeled the electroweak and strong forces under the single umbrella of the larger symmetry group SU(5). The expansion in the mathematical space meant that the total number of force-carrying particles was expanded to twenty-four.[23] New types of Higgs particles were also required to explain how the original unified symmetry had broken down.

The Glashow-Georgi model was an extremely elegant theory—so elegant that many believed it had to be right on aesthetic grounds alone.[24] Being a unified theory, it saw particles as manifestations of a single unified field—so, for example, quarks and leptons were related. Any kind of transformation could be viewed not as real change, but as the rotation of a single entity in the mathematical space. Parmenides had argued that change was impossible; for an entity to transform into something else, the thing which it had been must disappear, which was impossible. A unified theory resolved this paradox because it made everything one. However, there turned out to be a catch.

A consequence of the theory was that protons should be unstable, in the sense that its constituent quarks could be transformed—or rotated, in the terms of symmetry—into an electron or a neutrino. The expected lifetime of a proton would be longer than the length of the universe; but if you had enough protons—say, those contained in a really large tank of water—then a number should decay over the span of a year. In the 1980s, a number of experiments were set up to spot such decay events, but not a single one was observed. This came as a shock to a number of physicists. As Edward Farhi from MIT put it, "SU(5) was such a beautiful theory, everything fit into it perfectly—then it turned out not to be true."[25]

If the Standard Model was to be unified in a grand way, it seemed that something was missing. One of the most challenging aspects of the problem was the difference between the fermions, which make up matter, and the force-carrying bosons. As discussed above, these particles are as different as odd and even—the fermions have spins equal to an odd multiple of 1/2 and obey the Pauli exclusion principle, while bosons have integer spins and can share the same quantum state. One approach known as supersymmetry, which first became popular in the 1970s, asserted that fermions and bosons were not different because again they could be viewed as two aspects of the same thing.

What We Mean by Beauty

According to supersymmetry, every fermion has a partner boson and vice versa. The original hope was of course that the existing fermions and bosons would match with each other, thus reducing the number of fundamental particles. Unfortunately, this didn't work out, so instead new particles had to be invented. The naming convention was that the prefix "s" or suffix "ino" was added to the name of the original fermion or boson respectively. For example, an electron was paired with a boson called a selectron; quarks were paired with squarks; photons were paired with photinos; W particles with winos, and so on.

Supersymmetry therefore performed a similar trick to the SU(5) unification by assuming a higher-order mathematical symmetry that unified particles in one sense, at the expense of introducing many more of them. In the theory's original version, the selectron was predicted to have the same mass and charge as the electron, and it was thought to be similar for the rest. However, no such particle as a selectron was detected in experiments, and none of the other supersymmetric particles (or sparticles) appeared to exist either. It was therefore proposed again that the symmetry must have been broken by some process that granted the particles different masses, which were above the range of then-existing accelerators. In the simplest version of this theory, known as the Minimal Supersymmetric Standard Model (MSSM), the number of particles doubled and the number of free parameters increased from 20 to 125.

This unification therefore made the model more complicated, rather than less, which again seemed a rather unappealing feature. However, it was found that with the addition of supersymmetry, the unification of the various forces would all have happened at exactly the same energy, occurring about 10^{-39} seconds into the big bang. The unification energy corresponded to an energy of 10^{15} GeV, which is well above the range of any possible particle accelerator, and the theory also assumed that the model can be extrapolated to such high energies without any new effects coming into play. The famous graph produced by Ugo Amaldi, Wim der Boer, and Hermann Furstenau in 1991, which showed how the strength of the separate forces converged as temperature increased, seemed to be compelling evidence of supersymmetry—and beauty.[26] As John D. Barrow noted, "It is a simple symbol of the Universe's deep unity in the face of superficial diversity, which is what we mean by beauty."[27] Or, in Murray Gell-Mann's words, "when the mathematics is very simple . . . that's essentially what we mean by beauty or elegance."[28]

Many physicists strongly believed, and continue to believe, that supersymmetric particles will be detected at the LHC. Edward Witten said in 2003 that "many physicists do suspect that our present decade is the decade when supersymmetry will be discovered. Supersymmetry is a very big prediction; it would be interesting to delve into history and try to see any theory that ever made as big a prediction as that."[29] The

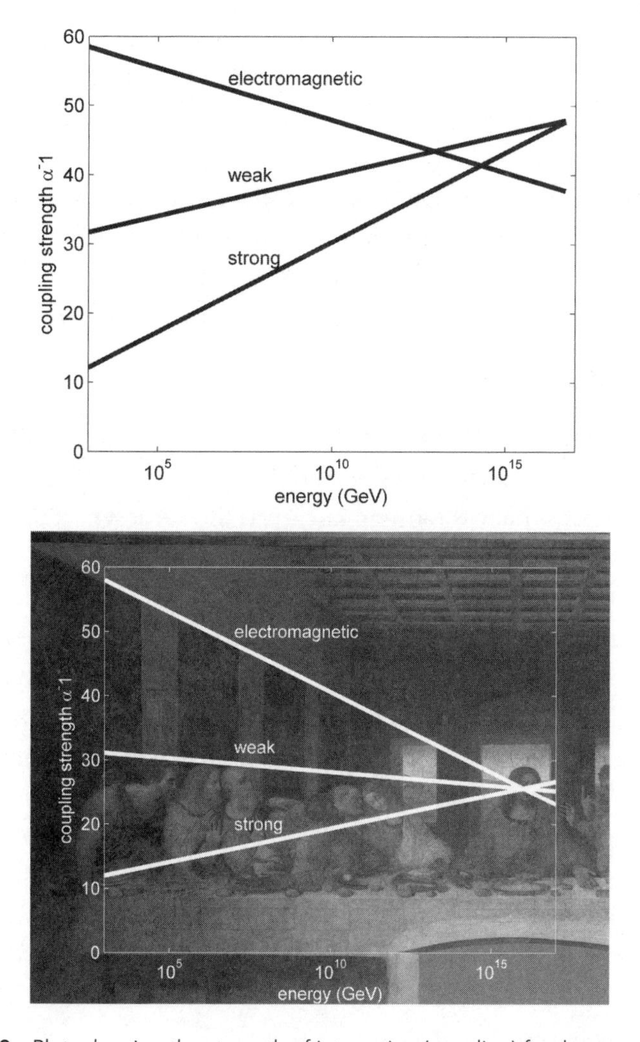

Figure 7.3 Plots showing the strength of interaction (coupling) for the strong, weak, and electromagnetic forces. The coupling is a dimensionless number which changes with energy because of the effects of virtual particles. The inverse of the coupling strength is predicted to vary linearly when plotted against energy on a logarithmic scale as shown. The top panel shows the case for the Standard Model, while the bottom panel shows the case where the effects of supersymmetric particles are included. Here the three forces converge to a single point, like the lines of perspective in a Renaissance painting, past which they would merge into a single force. As one commenter wrote, "I'd imagine the person who plotted it for the first time must have been completely ecstatic."[30] Note that data are only available at the low end of the energy scale; the assumption that the lines are perfectly straight even at very high energies is based only on theory. Source: Author and Leonardo da Vinci, *The Last Supper*, 1495–98. Santa Maria delle Grazie, Milan.

physicist Gordon Kane, who has spent some thirty years working on the theory, describes it as "wonderful, beautiful and unique."[31] According to physicist David Gross, the theory is "beautiful, natural, and unique. . . . Einstein would, if he had studied it, have loved it."[32] Theorist Michael Peskin calls it "the next step up toward the ultimate view of the world, where we make everything symmetric and beautiful."[33] But at the time of writing, no supersymmetric particles have made an appearance at the LHC or elsewhere. Their beauty is proving more elusive than anticipated.

String Theory

Even with supersymmetry, Grand Unified Theories still only united three of the four known forces. A true Theory of Everything also had to include gravity, which as Einstein knew was the hardest part of the problem. Reconciling gravity with particle physics was particularly important for understanding the first moments of the big bang, when the cosmos was the size of an atom and all of spacetime would have been subject to quantum fluctuations.

In 1968, the Italian physicist Gabriele Veneziano was studying the scattering properties of particles involved in strong-force interactions when he noticed that the data could be interpreted by modeling particles not as points, but as tiny strings oscillating at different frequencies. The mass of a particle corresponded through $E=mc^2$ to the frequency of the oscillation. Strings that oscillated more quickly had more energy and therefore greater mass. To match his experimental data, Veneziano assumed a string length smaller than 10^{-15} meters, or the size of a proton, and a string tension of fifteen tons.[34]

The technique was initially adopted as a calculating device rather than a model of reality, but a number of physicists found it fascinating in its own right and began to explore the idea in more detail. The string approach seemed to work well for the strong interaction, which behaves in any case rather like a rubber band, diminishing with distance below a certain scale. But as scientists tried to generalize it to other forces and particles, they ran into a number of problems. For example, the strings had to exist in spaces with extra spatial dimensions—twenty-five, to be precise. To explain the fact that the space we live in is apparently endowed with only three dimensions, theorists proposed that the extra dimensions were shrunken down to a scale where they had no noticeable effect except at the subatomic level (see Box 7.2). It was originally hoped that the geometry of this extra space could be described by a version of a sphere, but it was later realized that something more complicated was required to make the results consistent with the Standard Model—namely a Calabi-Yau manifold (not a car part, but a mathematical structure).[35]

Another problem was that strings could only model the force-carrying bosons, but not the fermions which actually make up matter. Not to be dissuaded, theorists including Pierre Ramond, André Neveu, and John Schwarz found in the early 1970s that if supersymmetry were adopted, then the expanded model could include both fermions and bosons. Everything was made of the same unified stuff. The strings still had to exist within a higher-dimensional space, but the total number was reduced to nine spatial dimensions plus time, for a total of ten.

In this superstring theory, the properties of particles, including quantum numbers such as spin, corresponded to the harmonics of the vibrating strings. A beautiful feature of the theory was that it explained why quantum numbers came in whole number multiples. In the same way that de Broglie showed that an electron orbit was limited to an integer number of wavelengths, thus quantizing the orbit energy, so the harmonics of a string were related by whole numbers. Particles, like the frets of a guitar, represented different notes in the tune of the cosmos. Strings could split or connect to form new particles, or the ends could join to form a closed loop.

One of these closed loops corresponded to something that hadn't been seen before: a spin-2 particle of zero mass. At first no one knew what to do with this, but John Schwarz and Joel Scherk soon noticed that the particle had the required properties for the hypothetical graviton, which carries the gravitational force.[36] Superstring theory was therefore a genuine candidate for a Theory of Everything. Schwarz later called this the first miracle of superstring theory. Bringing gravity into the picture implied a much smaller string length of no more than 10^{-35} m and a correspondingly enormous tension of 10^{39} tons. Nature, it seemed, was very highly strung.

Physicists really started to take notice in 1984 when Schwarz and Michael Green published a paper entitled "Anomaly Cancellations in Supersymmetric D=10 Gauge Theory and Superstring Theory," which claimed that superstring theory did not suffer the infinities which had long plagued quantum field theory (though see discussion in the next section).[37] This "second miracle" represented a major turning point (the third miracle had something to do with the mass of an electron not being infinity, which was indeed an excellent outcome). Superstring theory started to grow in popularity until it dominated most physics departments around the world. A 2001 article in the *CERN Courier* wrote, "it would not be an exaggeration to say that supersymmetry dominates high-energy physics theoretically and has the potential to dominate experimentally as well. . . . This belief is based on the aesthetic appeal of the theory, on some indirect evidence and on the fact that there is no theoretical alternative in sight."[38]

Supermodel

Actually, string theory is not the only exhibit on offer in the gallery of Theories of Everything. One alternative was presented by the Hawaii-based physicist/surfer Garrett Lisi in a 2007 paper called "An Exceptionally Simple Theory of Everything." His theory views particles as being the embodiment of symmetries in the mathematical group E8 (Figure 7.4), a distant relative of SU(2) and SU(3) which is renowned by mathematicians for its "beauty" and "uniqueness" on account of its amazing range of symmetries.[39] As Lisi wrote in *Scientific American*, "a unified theory provides a more aesthetically satisfying picture of how our universe operates. Many physicists share an intuition that, at the deepest level, all physical phenomena match the patterns of some beautiful mathematical structure."[40] The theory again predicts a range of new particles. Equally interesting are Lisi's ideas about work/life balance: "Surfing and snowboarding are what I do for fun—to get out and play in nature. We live in a beautiful universe, and I wish to enjoy it and understand it as best I can."[41]

Another approach, championed by a number of scientists including Lee Smolin and Carlo Rovelli, is called loop quantum gravity.[42] This theory views spacetime as being grainy, built up from networks of small lumps joined by links—rather like a very advanced knitting project. Clusters of links turn out to share the properties of quarks and other particles. This seems an interesting approach with a different aesthetic, but so far it has not delivered in terms of testable predictions.

In the eyes of the majority of physicists, however, string theory remains the current best hope for a Theory of Everything—and it has certainly attracted the most people and research funding. John Burnet wrote that, following the Pythagorean discovery of the numerical basis for musical harmony, Greek philosophy was "dominated by the notion of the perfectly tuned string." The same could be said today—though this might be beginning to change.

The main problem with string theory is that, like other such theories, so far it has not told us anything measurable about nature that we didn't already know. A number of physicists therefore believe that it has become detached from reality and goes against part of the spirit of science, which insists on experiments as a reality check. In a 1985 interview, Richard Feynman said, "I don't like that they're not calculating anything. I don't like that they don't check their ideas. I don't like that for anything that disagrees with an experiment, they cook up an explanation—a fix up to say 'Well, it still might be true.' . . . It doesn't look right."[43] In 2003, Sheldon Glashow told PBS, "The string theorists have a theory that appears to be consistent and is very beautiful, very complex . . . [but] there ain't no experiment that could be done nor is there any observation that could be made that would say, 'You guys are wrong.' The theory is safe, permanently safe. I ask you, is that a theory of physics or a philosophy?"[44] Along the same vein, physicist Robert Laughlin wrote, "String theory is immensely fun to think about because so many of its internal relationships are unexpectedly simple and beautiful. It has no practical utility, however, other than to sustain the myth of the ultimate theory. . . . Far from a wonderful technological hope for a greater tomorrow, it is instead the tragic consequence of an obsolete [reductionist] belief system."[45] In his book of the same name, Peter Woit borrowed the put-down "not even wrong" from Wolfgang Pauli to describe the theory's unfalsifiable nature.[46]

Indeed, most of the arguments for the existence of strings seem to be based on aesthetics. As Steven Weinberg, who has worked on the theory, admits, "string theory has been mostly driven by a sense of mathematical beauty and by a search for consistency."[47] John Schwarz felt that it was "too beautiful a mathematical structure to be completely irrelevant to nature."[48] According to his string theory colleague Leonard Susskind, David Gross felt that the theory "could not be wrong because its beautiful mathematics could not be accidental."[49] In particular, the theory is beautiful and consistent in the way that it aligns with the aesthetic standards of traditional science—which, as argued here, go back to the ancient Greeks.

Superstring theory is the perfect Pythagorean model—or supermodel—of the universe. It unifies all the known forces. It involves vibrating strings. It relates physical properties to whole numbers through the use of harmony. It is symmetric. It even takes place in a ten-dimensional space, the Pythagoreans' favorite number. So perhaps the theory is considered beautiful not so much because of the way it illuminates the world, but because of the way it slots neatly into an ancient, deeply ingrained, but largely unconscious and unexamined aesthetic template.

Note that, as with the Standard Model, while the underlying *ideas* of string theory are considered beautiful, their actual implementation is highly ornate and difficult for even specialists to understand. In the case of M-theory (see Box 7.2), no one even knows what it is. Its sense of beauty derives not from our world with its broken symmetries and ad hoc collections of particles and forces, but from an idealized primeval state where all was in perfect harmony; we live in a fallen, shattered version of the real thing. One can therefore argue that the attractiveness of string theory is over-rated.[50] Indeed, it seems that the more abstract and difficult a theory is, the more often the concept of beauty is evoked as a kind of risk-free defence. It is much easier to claim that a theory is beautiful, than to show that it actually works or makes sense.[51]

The confluence between modern physics and ancient philosophy is in a way rather wonderful, but it also has a less inspirational side. One can't help remembering that the Pythagoreans were a pseudo-religious cult with a strong sense of group conformity and all sorts of mystical ideas about the universe (they thought Pythagoras was descended from Apollo, they had numerous prohibitions including the eating of beans, and so on). Some observers of modern physics note that string theory may also have "mutated into something worryingly close to a religious cult," as science journalist Robert Matthews wrote in the *Financial Times*.[52]

As Lee Smolin documents in his book *The Trouble With Physics*, string theorists do demonstrate some rather Pythagorean behavior patterns. These include "tremendous self-confidence," an "unusually monolithic community," and in some cases "a sense of identification with the group, akin to identification with a religious faith or political platform."[53] He compares their behavior with the concept of "groupthink" as defined by the psychologist Irving Janis in the 1970s, which is characterized by the need to "preserve group harmony" above everything else.

As an example, Smolin cites the claim by string theorists that superstring theory is free of anomalies. This claim was first made in 1984 by Schwarz and Green, and it is routinely churned out; but, as Smolin

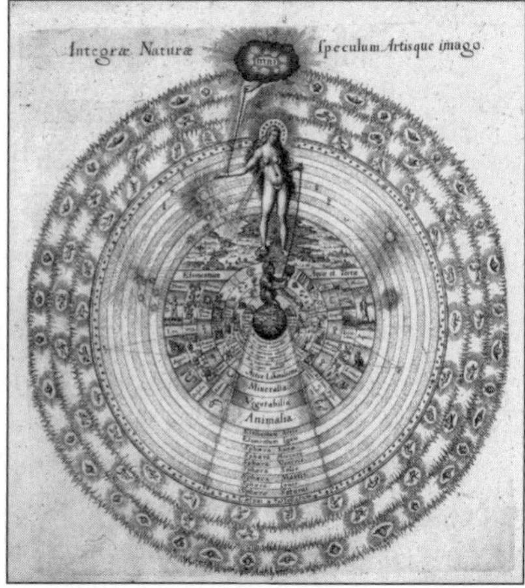

Figure 7.4 Left panel is a two-dimensional representation of the E8 group, showing the projection of the eight-dimensional "root" points which generate the group.[54] Right panel, for comparison, is "The Mirror of the Whole of Nature and the Image of Art" which represents an earlier Theory of Everything by Robert Fludd.[55] In this engraving from 1617, nature (the woman) is linked from her left hand to the ape at the center, who represents man's aping of nature through the arts and sciences. Her right hand links to God. The engraving also makes connections between the Sun, planets, alchemical transformations in the sublunary sphere, and so on. Source (left): American Institute of Mathematics. Source (right): Reprinted from Robert Fludd, *Philosophia sacra et vere christiana seu meteorologia cosmica* (Frankfurt: Francofurti prostat in officina Bryana, 1626).

points out, it simply isn't true.[56] Some of the calculations were shown to be finite, but the theory as a whole has never been shown to be free of infinities, especially when the nonlinear effects of spacetime curvature are taken into account. What was strange was that neither Smolin nor his colleagues "could recall ever having heard a string theorist point to it as an unsolved problem."[57] The desire for unity, it seems, has a shadow side: conformity.[58]

Box 7.2

M is for Magic?

The idea that strings exist in extra dimensions goes back to work in the 1920s by the German physicist Theodor Kaluza and the Swedish physicist Oskar Klein. According to the Kaluza-Klein theory, every point of four-dimensional spacetime has a miniscule circle attached to it that extends into the fifth dimension.

The new circle was too small to see or detect. But when Einstein's general theory of relativity was applied to this geometrical space, the extra dimension modified it in such a way that a new force appeared, which happened to look exactly like electromagnetism. Electric charge was identified as the momentum of a particle as it moved around the hidden circle, which had size of about 10^{-35} meters (a scale set by Planck's constant).

The physicist George Uhlenbeck wrote about the discovery, "I feel a kind of ecstasy! Now one understands the world."[59] Einstein, who was likewise looking for ways to unify electromagnetism with general relativity, was also initially impressed. Unfortunately, the theory demanded that the radius of the little circle remain constant at all times and points, which he noted was incompatible with the dynamic nature of general relativity.[60]

The theory made a comeback in the 1970s when string theorists revived it as a way to incorporate their higher-dimensional strings. The missing part of the strings was assumed to be curled up in the hidden extra dimensions. As skeptics pointed out, however, the addition of extra dimensions allowed the models to include many extra unmeasurable parameters, all of which could be adjusted to give the required answer.

Indeed, the inherent flexibility of strings means that there are several different versions of the theory available. For example, one theory suggests that what look like strings could actually be a manifestation of higher-dimensional membranes. According to Edward Witten from Princeton's Institute for Advanced Study, these theories are all different aspects of a single larger theory, known as M-theory, which still awaits discovery.[61] No one (except perhaps Witten) knows what the letter M stands for—suggestions have been membrane, magic, mystery, or an upside down letter W for Witten's last name—but it is reminiscent of the temple for Apollo at Delphi, whose walls were adorned with a number of statements, including for some reason a single letter E (or maybe it was an M on its side).[62]

Furthermore, variability of the details in which hidden dimensions are curled up lead to even more possible permutations—depending on how you count, 10^{500} or even more.[63] Together they make up what Leonard Susskind called the string "landscape," which is a term borrowed from evolutionary biology to describe the range of possible genetic variants.[64] For comparison, the number of particles in the observed universe is estimated to be about 10^{82}.

Many physicists believe that M-theory is the only serious candidate for a complete theory of the universe.[65] But not everyone is convinced. After all, the idea of attaching small circles to a geometric model has a pedigree that goes back to the ancient Greeks. They're called epicycles.

Conventional Beauty

If string theory has a central intellectual and inspirational figure, it would be Edward Witten.[66] He has been described as "the greatest theoretical physicist in the world" and "Einstein's true successor."[67] Stories abound about his amazing insight into the physical structure of the universe as well as his mathematical inventiveness, and his work has inspired whole new areas of mathematics. His decision to focus his attention on string theory in the 1980s effectively blessed the subject and made it respectable for others to follow suit. He has been awarded everything from the 1990 Fields Medal (the top honor in mathematics) to the University of Calabria's 2005 Pythagoras Award. (In *Not Even Wrong*, Peter Woit jokes that he was so awed by Witten that, after losing sight of him on a walk, it crossed his mind that "Witten was an extra-terrestrial being from a superior race who, since he thought no one was watching, had teleported back to his office."[68] Pythagoras was also said to be able to dart through space, but he did it using an arrow that had belonged to Apollo.)

In an interview for PBS's *The Elegant Universe* in 2003, Witten was asked what he meant by string theory being beautiful. He answered,

> Even before string theory, especially as physics developed in the 20th century, it turned out that the equations that really work in describing nature with the most generality and the greatest simplicity are very elegant and subtle. It's the kind of beauty that might be hard to explain to a person from a different walk of life who doesn't deal with science or math professionally. But the beauty of Einstein's equations, for example, is just as real to anyone who's experienced it as the beauty of music. We've learned in the 20th century that the equations that work have inner harmony. Now there must be skeptics out there who will tell you that these beautiful equations might have nothing to do with nature. That's possible, but it's uncanny that they are so graceful and that they capture so much of what we already know about physics while shedding so much light on theories that we already have.[69]

While the mathematics of it appears to be based on solid ground, supersymmetric string theory is clearly about more than equations. It is also based on strongly held and extremely powerful aesthetic principles—or Forms, to use Plato's expression—such as unity, harmony, and symmetry. As John D. Barrow notes, "Particle physicists are the most deeply Platonic because their entire subject is built upon a belief that the deepest workings of the world are based on symmetries."[70] The question, though, is whether the universe itself is—or once was—beautiful according to this scientific definition of "what we mean by beauty."

Consider for example the importance which is attributed to symmetry. Perhaps the most convincing and beautiful demonstration of the power of symmetry was Noether's discovery that conservation principles are the consequences of global symmetries. But while symmetry is a remarkably useful tool, not everything respects its constraints. The left-handed weak force, for example, is stubbornly asymmetric.

An early demonstration of the power of symmetry was its use to develop and simplify the equations for electromagnetism. But even electricity and magnetism are not completely symmetric with one another. It is possible to have electric point charges, but the equivalent for magnetism, known as a magnetic monopole, has never been detected. If you cut a bar magnet in half, you don't get one with a north pole and another with a south pole, but two bar magnets with two poles each.[71]

One of the greatest asymmetries, and one of the biggest unanswered questions in physics, is the imbalance in the universe between matter and antimatter. If the universe were exactly symmetric, then both should have been produced during the big bang in exactly equal amounts—in which case they either would have annihilated each other in a burst of energy, an embarrassing end to the universe before it even got going, like a rocket that explodes on launch; or else there would be a lot more anti-matter around today. Instead it seems that there was a small imbalance: a second after the big bang, for every billion particles of antimatter there were a billion and one particles of matter. The universe we see is what was left over—like a cancellation error—after this matter and antimatter collided and were annihilated.

Asymmetry is also a kind of marker for life. Louis Pasteur discovered in the mid-nineteenth century that organic chemicals such as amino acids can appear in two reflected versions. Artificially synthesized compounds will have a mix of both, but ones produced by living organisms always have the same handedness. He concluded that "life as manifested to us is a function of the asymmetry of the universe. . . . I can even imagine that all living species are primordially, in their structure, in their external forms, functions of cosmic asymmetry."[72] Just as symmetry is linked with unity and stasis, asymmetry is linked with plurality, change, and anything with a pulse. Of course, one can always assume that any observed asymmetry corresponds to the accidental breaking of some other perfect symmetry, which is the standard procedure in theoretical physics, but the deeper question is about aesthetics: are we inventing symmetries where they don't exist because we are convinced that they are beautiful according to conventional scientific standards? (Note again that these standards are not universal—beauty is at least partly in the eye of the beholder. The Zen Buddhist aesthetic of wabi-sabi, for example, is based on the idea that beauty is imperfect, impermanent, incomplete, and asymmetrical.[73])

Another fundamental cosmic asymmetry is that time only flows in one direction. The traditional test of reductionist theories is their ability to predict, to look into the future, to (in a sense) undo the asymmetry imposed by time. String theorists have proudly pointed out that the theory predicts gravity.[74] True—but the hard part about making predictions is that you are supposed to do them *before* you know the answer. As Niels Bohr is credited with saying, "Prediction is very difficult, especially about the future."

More relevant is the "very big prediction" of supersymmetric particles. But with the lack of confirmatory data from the LHC, squarks and selectrons are looking increasingly like the modern version of the counter-earth, which was invented by the Pythagoreans to rescue their equally beautiful ideas about the numerical harmony of the universe. Physicist George Smoot, whose work on the cosmic microwave background helped confirm the big bang theory, said in 2011 that "supersymmetry is an extremely beautiful model. It's got symmetry, it's super and it's been taught in Europe for decades as the correct model because it is so beautiful; but there's no experimental data to say that it is correct."[75] Joseph Lykken of Fermilab agrees that "it's a beautiful idea. It explains dark matter, it explains the Higgs boson, it explains some aspects of cosmology; but that doesn't mean it's right. It could be that this whole framework has some fundamental flaws and we have to start over again and figure out a new direction."[76] Of course, while the counter-earth hypothesis could be falsified through better telescopes, the same cannot be said of supersymmetry, since it may be that the particles are only created at energies higher than can be achieved by any accelerator.

In contrast, recent results at the LHC have confirmed the predicted existence of the Higgs boson (or at least something that looks a lot like it).[77] This discovery was a stunning vindication of the Standard Model's basic soundness, and backed up the idea that the acquisition of mass is related to a kind of cosmic phase change. It is the crowning achievement of a particle physics program stretching back many decades.

Of course, if the particle turns out to behave exactly as expected, then this will also freeze in place the aesthetic concerns and irregularities in the Standard Model that the LHC was partly meant to address. Phase changes are emergent properties that often produce intriguingly exact patterns but also act as a kind of barrier to further reduction—the Standard Model may end up being trapped in ice, with all its arbitrarily set parameters, and ungainly juxtaposition of different symmetry groups. Many physicists therefore hope that the discovered boson will have some unanticipated properties, since this would provide space for new theories to emerge (indeed, the possibility that the LHC finds the Higgs and nothing else has been described as a "nightmare scenario" for physics).[78] Personally, I will not be surprised if nature has a few tricks up its sleeve.[79]

Tinkertoy Symmetries

While researching this book, I encountered many statements about the beauty of modern physics, some of which are reproduced here. But I have to admit that most of them leave me a little cold. It isn't that I am a total philistine, or don't appreciate beautiful equations, or don't think the universe is beautiful—only that I find things like perfect unity and symmetry uninspiring. (And the idea of graduate students wandering around in T-shirts emblazoned with smug logos that purport to represent the structure of the universe somehow fails to give me the requisite aesthetic thrill.) Or perhaps the problem is that I'm too much of a purist.

In my first degree at university, I studied pure mathematics. It was possible to do this as part of either the science program or the arts program, and I chose the latter because it allowed me to take courses in a range of subjects such as philosophy, Shakespeare, art history, film studies, and so on. In mathematics, the equation that struck me the most was the Euler equation, which reads: $e^{i\pi}+1=0$. Here $e = 2.71828\ldots$ (etc.) is the number with the property that its logarithm is 1; while $\pi = 3.14159\ldots$ is the ratio of a circle's circumference to its diameter; and i is the square root of -1. The amazing thing about this equation is that it brings together these classic but apparently unrelated numbers, each of which plays an important role in completely different branches of mathematics, in a new and unexpected way—just as Einstein's $E=mc^2$ brings together energy and mass.

Like many mathematicians, I find the equation very beautiful as a piece of mathematics. It has both unity and symmetry. In fact, like U(1) in the Standard Model, it is based on the equation of a circle in the complex plane. For some reason, though, my appreciation of it depends on the layout of the equation—if it is written $e^{i\pi}=-1$ instead, it just looks like a calculation. Also, this type of beauty and elegance is somewhat narrow and specialized—it has something of the quality of a really good one-liner, like an advanced form of joke (brevity being the soul of wit).

It seems to me that physics is trying to achieve the same effect, but—with very few exceptions, such as $E=mc^2$—is finding much less success. I agree that aesthetic qualities such as unity and symmetry are useful, and indeed have helped to shape the history of science, but like any aesthetic guide they can become overrated. If architects always followed such simple rules, we would be living in very dull cities. If artists only cared about symmetry, then they would just draw different versions of geometric objects, and a portrait of E8 would be valued more highly than the *Mona Lisa*. When I saw Leonardo's *Portrait of an Unknown Woman* (also known as *La Belle Ferronnière*) at the National Gallery in London, the exhibition guide said that he had "so idealised her features that she may also be regarded as a perfect beauty based on ideas of divine geometry," but even there what struck me was how the depiction of the girl's personality seemed to swell up against those restraints.[80] It isn't the geometry that makes the painting beautiful; it is the tension between the geometry and what swims beneath. Beauty is itself an emergent phenomenon, which eludes tidy rationalization.

I have only sketched the outlines of the Standard Model and superstring theory here, so haven't been able to convey much sense of their complexity. My aim is to describe the aesthetics rather than the details of the physics. But these are supposed to be reductionist theories that gain their beauty from the simplicity of their concepts, if not their actual implementation. And even if the mathematics are hard or in some cases impossible to understand (as with the magical M-theory), the aesthetic principles that guide these theories are open to analysis and critique.

For example, it is striking that the multicolored quarks line up in neat patterns, like those in Figure 6.2, but nearly everything else about them is much less tidy, and I wouldn't interpret this all as a sign from the god of number. The Standard Model certainly impresses by its mathematical complexity and sophistication, and (more importantly) by the fact that it allows highly accurate simulations of subatomic experiments—not to mention the accurate prediction of new particles. It clearly represents a triumph of reductionist science. But in comparison to the richness of nature, the aesthetic pretensions of our models made me recall a segment in Martin Amis's 1995 book *The Information*, in which one writer criticizes another's novel for "its hand-me-down imagery, the almost endearing transparency of its little color schemes, its Tinkertoy symmetries."[81]

There is also another side to the mathematician's desire for simplicity, which non-mathematicians don't always get: it comes not just from a desire for beauty, but also from a kind of deep laziness, a desire for shortcuts. When the mathematician Carl Friedrich Gauss was a schoolchild, his teacher set the class a problem, which was to add up all the numbers from 1 to 100. The idea was probably to win the teacher some rest, but within minutes Gauss handed his slate in with just one number written on it: 5050. He had probably reasoned that the numbers from 1 to 100 could be grouped into 50 pairs: 1 and 100, 2 and 99, 3 and 98, and so on. Each pair adds to 101, so the total is 101 multiplied by 50, or 5050. It's a beautiful answer, but it's also a time-saver. (The fact that mathematicians are happy to come up with time-saving solutions for problems that no one else has actually thought of is just one of those paradoxes of human nature.)

The problem with the Standard Model, or supersymmetry, or string theory, is not that they fail to achieve perfect unity, harmony, symmetry, and so on. It's that they may be trying too hard—and the universe probably isn't like that anyway (unlike impatient mathematicians, it has all the time in the world). This wouldn't matter much if it weren't for the fact that—like characters stuck in a bad novel—other aspects of our lives and culture are being reshaped by the very same, rather corny aesthetic principles. In the next chapter, we will look at how the aesthetics of physics have affected both other sciences and our everyday lives, and argue that it is time to let go of old standards and open ourselves to different conceptions of beauty.

III

MATURATION

8

The Shadow World

Particle physicists, other physicists, and other scientists have been saying for some time that the product of our work is not merely arcane theories that we understand and others don't. It's not merely new devices, new medicines, new weapons. The product of our work is a world view that has led to the end of burning heretics and . . . to an understanding that we are not living in a world with a nymph in every brook and a dryad in every tree.

Steven Weinberg, *Facing Up: Science and its Cultural Adversaries*

The discoveries of science, the new rooms in this great house, have changed the way people think of things outside walls . . . It is my thesis that [these discoveries] do provide us with valid and relevant and greatly needed analogies to human problems lying outside the present domain of science or its present borderlands.

J. Robert Oppenheimer, *Science and the Common Understanding*

The economics profession went astray because economists, as a group, mistook beauty, clad in impressive-looking mathematics, for truth.

Paul Krugman, "How Did Economists Get It So Wrong?"

Perhaps the most perplexing problem faced by physics is that, while it does a reasonably good job of describing the energy and matter that we can actually see, it seems this visible matter only amounts to about 4 percent of the total. The rest of the cosmos is believed to be made up of mysterious quantities known as dark energy and dark matter, which exert peculiar gravity-type forces but are otherwise undetectable. In a way, it seems a good metaphor for life: we spend all our time analyzing and thinking about the small fraction of things that happen to be brightly lit while all the real action is going on somewhere else. This chapter looks at how our emphasis on finding a perfect model has affected our approach to a variety of scientific and social issues that have more influence over our lives than dark matter, but seem equally inscrutable—including the weather, our health, and the economy.

In Western culture, reason has long been associated—at least in metaphorical terms—with light, and ignorance with the dark. According to the classical scholar Francis M. Cornford, the Pythagoreans held that, "Light is the medium of truth and knowledge; it reveals the knowable aspect of Nature—the forms, surfaces, limits of objects that are confounded in the unlimited darkness of night."[1] In Plato's *Republic*, the Sun is described as the offspring of the Form of the Good, and the world of Forms is represented by light: "When [the soul] is firmly fixed on the domain where truth and reality shine resplendent it apprehends and knows them and appears to possess reason, but when it inclines to that region which is mingled with darkness, the world of becoming and passing away, it opines only and its edge is blunted, and it shifts its opinions hither and thither, and again seems as if it lacked reason."[2]

Plato's "allegory of the cave" compared the human race to prisoners in a dark cell whose view of the world outside is limited to shadows cast on a wall. Like those prisoners, the only way for a philosopher to break through and see "all things beautiful and right" is to throw off his metaphorical chains and escape into the sunlight. The aim of education was to lift the guardian class "from darkness to light" through training in mathematics and the other core subjects.

The mechanics of how light and sight actually worked remained somewhat fuzzy. How could our eyes gather information from the world around us and organize it all into a coherent image? One theory, which Aristotle favored, was that objects emitted some kind of signal that made us perceive them as a whole; for example, a table would place a little table icon in our field of view. Euclid took a more geometric approach and argued that the eye sends out light rays which travel in straight lines to their target.

Euclid's theory worked for bats—which "see" by sending out high-frequency squeaks and interpreting the echo—but it applied less well to humans. It was only in 1604 that the astronomer Kepler, who knew about lenses, provided the modern explanation—namely that the eye's lens focuses light upside down onto the retina, and the brain then interprets that information.[3]

In many respects, the scientific project can be viewed as a quest for better ways of seeing—both through better equipment to sense and better theories to make sense. It is the coupling of technology and brain power to extend our perception to greater distances and smaller scales, and to cast light on the unknown. Alexander Pope's epitaph for Newton read: "Nature, and Nature's Laws lay hid in Night / God said, Let Newton be! And All was Light."[4] Science during the Enlightenment did what it said on the label: spread the light of reason. Today, we have space telescopes and colliders to scrutinize the universe with ever greater levels of power and resolution; and advanced mathematics to enhance our abilities of analysis and discrimination.

It is therefore somewhat embarrassing to admit that 96 percent of the universe appears to be invisible. Something is out there in the dark and we don't know what it is.

Dark Matter

The existence of dark matter was deduced through observations first made in the 1930s that stars in our own galaxy were moving around too quickly to be contained by their own gravitational pull.[5] Other galaxies were also moving too fast relative to one another. Something seemed to be applying an extra gravitational force. Candidates included massive compact halo objects (MACHOs), such as black holes, or weakly interacting massive particles (WIMPs), which could be made of the famous supersymmetric particles, such as gluinos or squarks. Cosmologists now think that there aren't enough MACHOs around, so they see WIMPs as the more likely explanation. Another possibility is some generic new substance known as quintessence (from the Latin *quinta essentia*, or fifth element), which is an old name for ether. Whatever it is made of, dark matter is currently estimated to make up 22 percent of the total mass of the universe. It must have played an important role in the early universe, as clouds of dark matter helped to congregate visible matter into sufficiently high concentrations to build stars and galaxies.

While dark matter is holding the galaxies together, it was discovered in the late 1990s that another force appears to be pushing the universe as a whole apart, so its rate of expansion is accelerating. Under Einstein's general relativity, the cause appears to be an energy density, which translates through $E=mc^2$ to a mass equal to 74 percent of the universe total. To quote the Marquis de Condorcet's remarks on the phlogiston theory of fire (which could be thought of as the dark energy of the eighteenth century), it seems that the universe is "impelled by forces that give it a direction contrary to that of gravity."[6]

Together, dark matter and dark energy are believed to account for 96 percent of the total mass-energy content of the universe. That leaves a mere 4 percent in the form of visible matter, of the sort accounted for by the Standard Model. As the astronomer Robert Kirshner put it, "The universe may be like Los Angeles. It's one-third substance and two-thirds energy."[7]

An alternative explanation, of course, is that our models are wrong. Perhaps for example Newton's law of gravity doesn't hold over extremely large scales, as physicist Mordehai Milgrom proposed in 1983.[8] However Lee Smolin notes that such ideas are considered "too scary to contemplate," and indeed we seem to have more faith in our ability to compute the universe, than to actually see it.[9]

In any case, this darkness doesn't seem like too much of a problem, since the fact that we can't detect it means it probably won't disturb our everyday lives. We can excuse physicists for concentrating only on conventional forms of matter—the sort that we are likely to interact with. No one would accuse them of being excessively blinkered or blinded by ideology. However, there are many other areas of science where a narrow focus—and an emphasis on aesthetics—may be limiting us to a similarly restricted view of the world.

It is often said by particle physicists that everything in the universe is ultimately made up of elementary particles. So, to understand the universe, we need to understand these fundamental building blocks. Everything relies on our model of subatomic physics, which is why it so important and beautiful. As string theorist Luboš Motl wrote, "If someone is not impressed by the fact that a formula . . . can explain a large number of physical situations, including chemistry and animals, as well as the sunset, she can never understand why the physicists think that string theory is beautiful. From this perspective, string theory is the ultimate achievement of reductionism."[10] Particle physicist Brian Cox said that quantum field theory "underpins our understanding of modern technology, and the way chemistry works, the way that biological systems work—it's all there. This is the theory that describes it all."[11] According to Steven Weinberg, any question ultimately traces back "through many intermediary steps to the same source: the Standard Model of elementary particles."[12]

This insistence on the importance of particle physics seems rather forced. In a trivial way, it's true that everything depends on quarks, electrons, and so on. But historically, the most actionable piece of information to have come out of particle physics was that uranium is explosive—and that discovery didn't require an accelerator. Engineers in the Victorian era didn't need to know the mass of an electron to exploit electricity for various purposes (including powering the cathode ray tubes which led to the discovery of the electron). The Standard Model doesn't rate highly on the must-read list for most biologists.[13] For an architect to design a better building, they could spend their lives studying the properties of bricks—or, for that matter, quarks—but it probably wouldn't be the best way forward. (Actually, Vitruvius's ancient text on architecture did include a chapter on "The Primordial Substance According To The Physicists," in which he describes atoms as the ultimate in bricks: "they cannot be harmed, nor are they susceptible of dissolution, nor can they be cut up into parts, but throughout time eternal they forever retain an infinite solidity."[14])

The idea that everything flows down from physics and mathematics is an idea that goes back to the ancient Greeks, but it is slightly misleading. It isn't information that flows down, but something even more potent.

It is an aesthetics—a way of seeing and understanding. And it affects everything from the way we handle the environment to the way we treat disease, to the way we run the economy.

Perfect Model

The whole point of reductive science is to be able to predict and control nature. I became interested in the relationship between science and aesthetics while working and giving talks on predictive models in different areas, including weather, biology, and economics.[15] These models lack the perfect symmetries and unities which can be found, or are sought, in subatomic physics, but they are still often treated—perhaps unconsciously—as an essentially perfect representation of reality, based on beautiful laws. Unfortunately, while the Standard Model can claim excellent accuracy and agreement with experimental data, the same cannot be said for these more applied models, which are used to predict things that directly affect our lives. The rest of this chapter will therefore pull itself away from the timeless and ethereal beauty of theoretical physics to explore how its influence has shaped our approach to the messier task of modeling the world. (Because I have written about other aspects of these topics elsewhere, the treatment is somewhat compressed and concentrates on tracing the role of aesthetics in their development.)

I first came across the expression "perfect model" when I was studying weather forecasting models as part of my graduate program. Following the demise of the Superconducting Super Collider in 1995, I had returned to university to do a doctorate in applied mathematics. My thesis topic involved determining the causes of forecast error in weather models—a question of interest not just for people planning their weekend, but for industries such as transport and agriculture.

Computerized weather forecasting was pioneered by the mathematician John von Neumann. One of the most talented and productive mathematicians of all time, he played a major role in the Manhattan Project, the development of computers, and later in setting U.S. scientific and military policy. During the 1940s he was working on simulations of nuclear weapon explosions when he realized that the same techniques from fluid dynamics could be used to forecast the weather.[16] If the weather could he predicted, he argued, then perhaps it could also be controlled—and even used as a weapon of war.[17] A hurricane, for example, releases about ten times the total yield of the Little Boy atomic bomb every second. (Von Neumann also later championed the doctrine of Mutually Assured Destruction, however, so presumably threats of American droughts and Russian winters would have canceled each other out.) He also believed the models could be used to predict climate and appointed Jule Charney to lead that effort.

A weather forecast model works by dividing the Earth's atmosphere up into a grid and using equations based on Newtonian dynamics to compute how the system will evolve over time. The first step in making a forecast is to measure or estimate quantities such as wind speed and direction, temperature, air pressure, humidity, and so on for each of the grid points. These numbers define an initial state for the atmospheric system, which has an inherent uncertainty due to measurement error. It is then fed into a computer, which runs the model to produce a forecast for how the system will evolve.

Today, supercomputers are about a million times faster than the ENIAC computer used by von Neumann. There have also been huge advances in the quality of weather observation—for example through the use of satellites—and modeling expertise. But despite these developments, the accuracy of forecasts has not improved as much as anticipated. Predictions of precipitation, for example, are only useful for a few days at most, and extreme events such as storms continue to frustrate our desire for accurate prediction. Controlling hurricanes is right out.

In the 1960s, the mathematician and meteorologist Ed Lorenz from MIT suggested a reason for the weather's unpredictability. He proposed that the atmosphere is very sensitive to small perturbations—so sensitive, in fact, that the flap of a butterfly's wings is enough to almost magically cause a storm a few weeks later on the other side of the world. The fact that "we do not know exactly how many butterflies there are, nor where they are all located, let alone which ones are flapping their wings at any instant" obviously complicated the forecaster's job.[18] On the bright side, the "butterfly effect" did help explain why forecasts are so unreliable.[19] It also didn't affect longer-term climate predictions, which were concerned with the average state of the weather rather than its short-term evolution.

Lorenz's work was part of the chaos/complexity revolution that began in the late 1960s, and that eventually led to a new scientific aesthetic and new methods of analyzing natural systems. But in weather forecasting, chaos theory was co-opted to reinforce the status quo, with minor modifications. According to researchers, the existence of chaos didn't imply that there was no point in making conventional forecasts—only that we needed many more of them.

In the 1990s, weather centers in Europe and the U.S. developed a new technique called ensemble forecasting. While weather models are based on Newtonian dynamics, the inspiration for ensemble forecasting can be "traced back to the Manhattan Project and ultimately to quantum mechanics," as one study notes.[20] In quantum physics, the ensemble interpretation, which was favored by Einstein, says that computations only give information about an "ensemble" of particles rather than individual particles. Similarly, the inherent uncertainty of weather observations means that the model should be applied not to a single estimate of the current weather, but to a number of possible initial states. This would yield an ensemble of forecasts, which would in some sense capture the truth within its net. As another history of the topic put it, "many numerical 'butterflies' wings' were to be flapped in the model's atmosphere," and the variability would give an idea of the uncertainty due to chaos.[21]

A problem with this approach is that weather models contain many thousands of variables, including the temperature, pressure, and so on at each grid point, so it is impossible to account for errors in all of them. Ensemble forecasters therefore adopted a probabilistic technique known as the Monte Carlo method, originated by Stanislaw Ulam at Los Alamos National Laboratory, again while working on the nuclear weapons program. This involves making random draws from a larger distribution and using those to infer the properties of the rest. Instead of running the model from thousands of different starting points, forecasters could randomly select just twenty or so.

When scientists started using this technique, they found that instead of rapidly diverging, as expected by chaos theory, the different forecasts remained closely clustered together. Rather than take this as a clue, they spent a lot of time contriving methods to produce initial states that gave a larger spread. The assumption throughout was that error was caused by the measurements of the initial state, coupled with the effects of chaos. To make calculations easier, theorists adopted the "perfect model" assumption, which stated that to all practical purposes the model could stand as a flawless representation of reality.[22] Later versions incorporated perturbations to some model parameters, but assumed that the basic structure of the model (i.e., the form of the equations) was correct. Unlike observation error, model error does not lend itself easily to the ensemble approach, because if the equations have the wrong form it is not clear what type of perturbation is appropriate.

The "perfect model" expression stuck in my mind. Models of the atmospheric system are not particularly elegant; indeed, from a mathematical point of view they are clunky, unwieldy contraptions. However, it is important again to distinguish between the end result and the underlying ideas, or ideals. Ptolemy's model of the universe was complicated, but based on beautiful circles. The Standard Model is baroque, but at its heart are beautiful spheres. M-theory is apparently beyond human comprehension, but everyone can appreciate the harmony of strings. In the same way, atmospheric models incorporate a kind of secondhand beauty in the sense that they are derived from the "laws" of fluid flow, which are in large part empirical but based to a degree on Newtonian mechanics and concepts such as conservation laws.[23]

Coming as I did from a background working on engineering projects, I knew that mathematical models were prone to error even in highly controlled environments. My research indicated that the butterfly effect was a myth. When butterflies flap their wings, the perturbation to the atmosphere is rapidly damped out (to get an idea, try flapping your hand in the air a foot in front of your face—does it feel like the effect is growing exponentially?). Sensitivity to initial conditions is certainly a feature of some mathematical equations, but as Stephen Wolfram points out, "none of those typically investigated have any close connection to realistic descriptions of fluid flow."[24] Weather models are not particularly sensitive to even large perturbations of the sort typically created by observation errors.[25] The fact that this undemonstrated theory has been evoked for decades as the default explanation for prediction error in many fields tells us more about the sociology of science—and the attraction of a pretty idea—than it does about reality.[26]

The real cause of forecast inaccuracy was something much less aesthetically and professionally appealing: model error. Something was out there in the atmosphere, pushing forecasts astray, and modelers didn't know what it was.

Dark Clouds

What is the source of this atmospheric version of dark matter or dark energy? To quote Lord Kelvin in his 1900 lecture, what "clouds" are obscuring the "beauty and clearness of the dynamical theory"?[27] What is blocking the light of reason?

One of the most likely candidates is just that—clouds. These are formed when water vapor condenses around minute particles in the atmosphere, but the process is highly nonlinear and involves all different scales from the large to the microscopic. Weather models try to compute the average level of cloudiness for an area from factors such as water vapor and air pressure, but they are poor at predicting how clouds will form, dissipate, or cause precipitation. The reason is that clouds are best viewed as an emergent property of the atmosphere, of the type which is not well-described by equations (see Box 8.1). Scientists know a lot about particles, air, and water, but they still can't accurately model a cloud. And knowing about quarks is really no help at all. Refining the resolution of the model, for example by reducing the grid size, often leads to only marginal improvements.[28]

Another troublesome example of emergent behavior is turbulent flow, which characterizes much of what we call the weather (the reason planes fly at high altitudes is that they want to avoid it). The British physicist Horace Lamb said in a 1932 speech that "I am an old man now, and when I die and go to heaven there are two matters on which I hope for enlightenment. One is quantum electrodynamics, and the other is the turbulent motion of fluids. And about the former I am rather optimistic."[29] Again, turbulent flow involves dynamics over all scales, cannot be reduced to Newtonian equations, and continues to resist the attentions of modelers.

Apart from emergent phenomena, a second property of the atmosphere that makes it difficult to model is that its dynamics are dominated by powerful feedback loops. Positive feedback occurs when a signal is amplified by the dynamics of the system; negative feedback occurs when the signal tends to be damped out. An example, again, is the effect of clouds on temperature. When the temperature is high, water evaporates, which increases the water vapor content in the atmosphere, which encourages the formation of clouds, which cools the atmosphere (negative feedback on heat)—except at night when cloud formation does the opposite (positive feedback on heat).

Because these feedback loops are in a state of dynamic opposing tension, they tend to partly cancel out and are difficult to accurately measure or simulate. As Heraclitus said, "Nature loves to hide." Their delicate balance also tends to make models unstable. The problem is not chaos, which is sensitivity to the initial state, but model error, caused by sensitivity to small changes in the model equations.

As models grow in size and complexity, two things happen. The first is that the dimension of the model increases. String theory sits in ten dimensions, but weather models have tens of thousands of dimensions, consisting of the values of quantities such as temperature and pressure at each grid point, and only some of these can be measured. Secondly, the number of adjustable parameters also increases along with model complexity. This means that the model can be adjusted to fit past data, but it doesn't get much better at making predictions about the future.[30]

The Standard Climate Model

On the surface, weather models have all the properties of a good scientific model. In aesthetic terms, they have the look and feel of hard science. They take an atomistic, reductionist approach by dividing the atmosphere into a fine grid of separate elements. They use deterministic, Newtonian-style equations to calculate the flow of air, water, and heat. They incorporate Democritian randomness—all things are "the fruit of

Box 8.1

Universal Constructor

In the early 1950s, the mathematician John von Neumann began experimenting with a primitive kind of automaton that could reproduce itself—and even evolve.

Von Neumann's cellular automata lived in a mathematical space consisting of a two-dimensional grid. Each cell of the grid could have twenty-nine different colors. As the system stepped forward in time, it progressed according to a set of rules, with the state of a cell in the next time-step depending on its four immediately adjacent neighbors. Von Neumann's aim was to mathematically prove that a large enough configuration had the potential to replicate—that is, to generate copies of itself. He believed that any such object had to be very complicated, consisting of around 200,000 cells.

This universal constructor, as he called it, was only an abstract set of instructions. Von Neumann originally developed the idea using a pencil and graph paper. But he thought it was a short step away from designing real, self-replicating machines that could carry out tasks that humans were ill-suited for, such as mining on the Moon. Machines could even evolve over time if mutations to the instructions were allowed to appear. The self-replicative ability of the program foreshadowed the later discovery of DNA.

In 1968, the Cambridge University mathematician John Conway resumed the search for a universal constructor. He eventually discovered a simple set of laws, later publicized by Martin Gardner in *Scientific American*, known as the Game of Life.[31] The game begins with some initial configuration of black counters, like a game of checkers on an infinitely large board. Each counter has eight adjacent cells, including diagonals. At each iteration, if two or three cells are occupied by neighbors, the counter survives; otherwise it is removed. These deaths are balanced by births in any empty cell that has exactly three neighbors. The fate of the system depends on the starting configuration; and for some, the screen suddenly becomes alive with an ever-shifting population of weirdly lifelike shapes.

The mathematician Stephen Wolfram classified the behavior of cellular automata in terms of four classes: (I) all activity dies out; (II) stable or periodic patterns; (III) random; and (IV) situated on the border between order and pure randomness, with a degree of structure but no enduring pattern. The most interesting were the latter group, which included the Game of Life. These showed what Wolfram called computational irreducibility: their behavior could not be predicted in advance using any set of equations.[32] Their properties were emergent features that could only be understood by running the system itself.

The existence of emergent properties means that reductionism fails. As Aristotle said in *Metaphysics*, the whole is more than the sum of its parts.

chance and necessity"—and have their intellectual roots in nuclear physics and quantum theory. If you hung them on the wall, they would look just fine with the rest of the scientific furniture. They also do a perfectly respectable job of predicting the weather a few days ahead (although ensemble forecasts have proved quite hard to commercialize). The problem comes when stripped-down versions of the same models are used to make medium- or long-term predictions—something for which they are clearly unsuitable.

In 2010, after a series of forecasts that included a "barbeque summer" which failed to ignite and a "mild winter" which became the coldest in thirty-one years, the UK Met Office decided to cancel public dissemination of its seasonal predictions.[33] Newspapers quickly switched to independent forecasters for 2011–12, but their "big freeze" and "Siberian December" turned out rather warm.[34] It is often said, without evidence, that long-term climate change is easier to predict than seasonal fluctuations, but this is simply wishful thinking.[35] As anyone who has tried to predict the evolution of a complex system—and that includes most business leaders—will agree, the default position is that accurate forecasts are impossible unless proven otherwise; and things don't get better the further out into the future you project, where the effects of long-term dynamics and feedbacks are famously difficult to foresee. The climate is always changing, and if models such as those used by the Met Office can't predict the weather over shorter time periods, then there is no reason to believe that similar models can predict what will happen to the evolving climate decades in the future, or accurately simulate its response to perturbations such as those caused by carbon emissions. This is illustrated by the lack of progress over the last few decades in reducing predictive uncertainty.

The first meeting to estimate the effect of carbon dioxide on climate was organized by von Neumann's appointee, Jule Charney, at Cape Cod in 1979. On the first day a model was presented which said that the warming due to a doubling of carbon dioxide would be 2 degrees Celsius. On the next day, another model was presented which said it would be 4 degrees. At the end of the conference, Charney split the difference and added a fudge factor to give a predicted range of 1.5 to 4.5 degrees Celsius, or 3 degrees plus or minus 50 percent.[36] This so-called canonical range has remained essentially unchanged ever since, which is rather strange given the huge advances in computing, meteorological technology, and models—and indeed the increasing evidence that warming is happening.[37] The most recent 2007 report from the Intergovernmental Panel on Climate Change (IPCC), for example, says that the answer "is very likely larger than 1.5°C" and "is likely to lie in the range 2°C to 4.5°C, with a most likely value of about 3°C," which sounds a lot like 3 degrees plus or minus 50 percent.[38] If Charney had chosen 2 to 6 degrees, then one can't help but suspect that would now be the range as well.

In fact the true uncertainty is even larger. As the climate scientist Stephen Schneider noted on his website in 2005, "despite the relative stability of the 1.5 to 4.5°C climate sensitivity estimate that has appeared in the IPCC's climate assessments for two decades now, more research has actually increased uncertainties!"[39] One Oxford University–led study showed that small changes in the model, which affected primarily the modeling of clouds, led to a projected warming of anywhere from nothing to 11.5 degrees.[40] These results, which do not account for structural errors in the form of the equations, are consistent with the idea that models are unstable and sensitive to small changes. At the same time, models are not good at reproducing extreme, suddenly evolving climate change events of the sort seen in the climate record (they are unstable in the wrong way).[41] As the pioneer climate scientist Syukuro Manabe put it, "Uncertainty keeps increasing with the more research money they put in. . . . It hasn't gotten any better than when I started forty years ago."[42] The official message, however, remains that models can be depended on for probabilistic predictions.

Science normally progresses when a new theory comes along that makes better predictions than the existing one. Einstein's relativity was accepted because it made more accurate predictions than Newtonian mechanics. But an inherent property of complex systems like the atmosphere is that they have evolved in such a way that they resist prediction. The emergent properties that characterize the climate, and the complex feedback loops that make it reasonably stable, also make it hard to predict. (Note that modeling and forecasting are separate but related activities—we can build complicated models of all sorts of processes, but models with predictive power are much rarer. Simple models are usually better for forecasting purposes than complicated models.[43]) If a better model doesn't emerge, then scientists don't have a mechanism by which to proceed—apart from aesthetic choice or funding, which are often related. Rather than admit failure, they often slide into groupthink, cohere around group decisions, and invent complicated explanations for their inability to make useful predictions.[44] Of course, scientists are far from unique in exhibiting this typically human behavior; they just aren't as immune as we would like to think.

Actually, climate models do have an unsung competitor in the form of a statistical method known as time-series modeling. For example, neural network models set up a network of artificial "neurons" that learn to detect patterns in past data. Because they do not attempt to reproduce the complex dynamics of the system, they are much smaller than traditional climate models and can run on a PC rather than a supercomputer. A 2011 study showed that, for a limited set of historical data, a neural network model outperformed a conventional climate model, while a combination of a time-series model with a conventional model led to an improvement of 18 percent in forecast accuracy over a ten-year period. Such time-series models are particularly good at spotting local variations, which tend to elude traditional models but are very relevant for policymakers.[45]

One drawback is that, because they are not based on a mechanistic explanation, time-series models cannot test (or pretend to test) hypothetical cases, such as the effect of high or low emissions. But if we don't have faith in the climate model predictions in the first place, then there is no reason to believe that their projections for different cases are accurate either. I would venture that the main reason why these statistical time-series techniques have not been incorporated for use in official climate forecasts is related to aesthetics—a topic which I doubt comes up much at meetings of the IPCC. According to an article in *Climatic Change*, the prevailing style of "good science" in climate modeling is "deterministic reductionist," which favors hard equations based on mechanistic analysis.[46] Methods such as neural networks can ultimately be expressed as a set of equations, but if you write them out they seem haphazard and strange—they look like a hack. But perhaps that ugliness is just an expression of the fact that the system under analysis is not easily reconciled with mathematics.

Uncertainty Principle

While ensemble forecasters have drawn inspiration from quantum physics, the real but unlearned lesson or metaphor from quantum theory is that we need an uncertainty principle for prediction. After all, if you study a system over a period of decades using instruments of ever finer resolution and yet the picture only gets blurrier, then this is a sign that you are bumping up against a fundamental limit. In a sense, this situation is even worse than that in quantum physics, because we cannot put a specific limit on the uncertainty. One can assign percentage probabilities for particular outcomes based on model simulations, but if the structure of a model is wrong, then perturbing it in some way only tells you about the sensitivity of the model, not the system itself. The only way to estimate the uncertainty for such a model is to compile statistics that compare forecasts with observations and hope that the future will resemble the past. As seen below, the problem is not unique to the climate; it also applies to other complex organic systems.

Of course, the fact that we cannot accurately predict the future climate does not imply that carbon dioxide emissions are safe or that we should stop analyzing the climate system. There is plenty of evidence—based not on forecasts, but on statistical data analysis—that the climate is both warming and becoming less stable as the result of human influence.[47] Climate change is only one aspect of a more complex environmental crisis, which also includes symptoms such as species loss, ocean damage, resource shortages, and so on. The process of modeling a system is an excellent way to learn how it works and appreciate its complexity, and mathematical modelers and climate scientists have helped draw our attention to the potential dangers caused by emissions. The biggest challenge to the environmental movement is not the failures of modelers, but the facts that proved fossil reserves in the ground represent tens of trillions of dollars in potential assets; the task of extracting them is one of the most profitable businesses on the planet; and some of the profits are devoted to lobbying and propaganda. Past a certain point, though, excessive reliance on models can become part of the problem. Our obsessive focus on trying to calculate the details of global warming has become a distraction.

The whole point of mathematical modeling is to stand back from the object under analysis and to treat it objectively. However, climate models incorporate many subjective choices, because the form of the equations and the values of the parameters cannot be determined from data alone (if the system is incomputable, then no correct set of equations even exists).[48] The insistence on objectivity is therefore something of an artifice. And paradoxically, what we probably need is *more* subjectivity and emotional engagement, rather than less. If as a society we wish to protect the environment, then distancing ourselves from it—and promoting the idea that we live in a closed, deterministic, machine-like world—may not be the best response. To close the gap, art is as important as science (an example is the organization Cape Farewell, which was founded in 2001 by UK artist David Buckland to bring together artists, scientists, and communicators in order to explore the effects of climate change).[49]

The pretense of prediction is also damaging because, in the Western psyche, the ability to predict is entwined with the ability to control. The idea that we can predict the climate but cannot fix it therefore comes across as weak and somehow contradictory. Criticizing climate models may be interpreted as giving ammunition to climate change "deniers"—which is one reason for the lack of engagement with this topic—but the poll numbers are in and the current approach is not working. While there has certainly been much progress in areas such as alternative energy and sustainable building, public concern about climate change has declined in most industrialized countries in recent years.[50] In the United States, belief that fossil fuel consumption will cause the climate to change has gone down, according to Harris, from 75 percent in 2001 to 44 percent in 2011—a truly staggering collapse that threatens the gains of the environmental movement.[51] The oracles no longer hold our attention. This is not just a communication or public relations issue, but a sign that the climate argument needs a different approach that is not propped up by scientific hubris, reliant on traditional ideas of predict and control, or slave to a defunct aesthetic.[52]

Rather than view climate change as a kind of elaborate physics problem, it would be more realistic to adopt a medical analogy. Carbon dioxide in the atmosphere is like glucose in a person's bloodstream. When maintained at the correct low levels, it is vital for survival, but if the concentration is doubled it leads to problems (diabetes for a person, warming for the climate system). In the case of our planet, this is a new condition; we don't know the prognosis, but the symptoms are troubling (for us, at least—the Earth has survived much worse). Like a patient with serious health worries, we don't have time for more research reports.

Emergent properties and interlocking feedback loops impose as great a limit on our ability to calculate the future as any quantum effect, only on a much larger scale.[53] This is not a cop-out; it's reality. Perhaps if we had listened more closely to the lessons of the atom rather than paper over them with the Copenhagen interpretation, we would be more open to that kind of uncertainty today. If our scientific aesthetic acknowledged the beauty of paradox and incomputable complexity, we might—paradoxically—face a less uncertain future.

Kepler aimed to show that "the machine of the universe" is "similar to a clock."[54] Rather than upgrading our models or computers, we need to change our aesthetic—from seeing the world as a machine to seeing it as a living system. Otherwise, our wait for an accurate Model of Everything for climate change is likely to continue until we can get the answer by looking out the window.

The Atoms of Biology

While meteorology and climate science are explicitly modeled on physics, life sciences such as biology and ecology have generally taken a more empirical route—perhaps because living systems exhibit rather less of the formal unity and symmetry beloved by physicists. However, these areas too have been influenced by the dominant scientific aesthetic with implications for the way we treat disease.

The success of atomism in physics meant that scientists soon began to look for an equivalent in other areas of science, including biology. In the late nineteenth century, biologists began to consider the cell as a kind of living atom. In Gilbert and Sullivan's *Mikado*, one character traces his ancestry to "a protoplasmal primordial atomic globule."[55] In 1925, the philosopher Alfred North Whitehead wrote that "the living cell is to biology what the electron and the proton are to physics."[56]

Unfortunately, cells lacked the pleasing integrity and uniformity of fundamental particles, and they did little to explain the mechanics of inheritance. It had been known since the mid-nineteenth century that certain traits were passed in a discrete fashion between generations. For example, Gregor Mendel showed that if a pea plant is descended from one parent with wrinkly seeds and one with smooth seeds, then its seeds will be wrinkly or smooth (rather than a mix of the two) with a probability that conformed to consistent mathematical ratios, reminiscent of the neat proportions that inspired the periodic table of the elements.[57] He concluded that, for any particular trait, each parent plant had two "factors"—later known as genes—and each would contribute one to the next generation.

Mendel's theory was mostly ignored until the early twentieth century, when experiments showed that organisms including fruit flies also inherited traits such as eye color in a Mendellian fashion. Furthermore, Hermann Müller found that fruit flies experienced a background rate of genetic mutations which could be accelerated by exposure to X-rays or chemicals but was always present.[58] Genes therefore provided a mechanistic explanation for evolution, which led to the theory known as Neo-Darwinism. This was a crossbreed between Darwinian evolution, which was dynamic, and Mendel's genes, which were apparently static. During sexual reproduction, the DNA from the parents is mixed and matched, so we inherit qualities from each. Random mutations of these genes lead to variety between individuals. Darwin's concept of natural selection acts as the mechanism that drives evolution.

During and after the Second World War, Neo-Darwinism became part of the ideological conflict between capitalism and communism. Capitalist countries favored Neo-Darwinism and its emphasis on competition; Russia and its communist allies preferred the unorthodox methods of agronomist Trofim Lysenko, who downplayed the role of heredity and argued for the non-genetic inheritance of acquired characteristics. The ensuing collapse of Soviet crop yields seemed to settle the matter. As is discussed below, Neo-Darwinism has continued to be influential in areas from economics to cosmology.

The Central Dogma

While the idea of genes was gaining believers, no one knew where these genes were or how they worked. In his 1944 book *What Is Life?*, Erwin Schrödinger suggested that genetic information was transmitted through some kind of code.[59] Cracking the code would be analogous to cracking the atom, and other physicists began to take note. Paul Dirac noted that genetics "is the most fundamental part of biology" and there are "laws governing the way in which one inherits characters from one's parents."[60] Attention soon focused on the substance known as deoxyribonucleic acid, or DNA, which appeared to be the "master molecule" in the chromosomes responsible for the transmission of genetic traits.

After Rutherford's death in 1937, his position as head of the Cavendish laboratory in Cambridge was taken by another colonial expat, the Australian William Lawrence Bragg. At the age of five, Bragg broke his arm after falling from his tricycle. His father, who had learned of the recently discovered X-rays, used them to image his arm—the first recorded use of the technique in Australia. Bragg would go on to become an expert in X-rays himself. He developed a technique that made it possible to infer the structure of crystals by studying the way in which they diffracted X-rays. X-ray crystallography soon proved indispensable in areas such as chemistry, material science, mineralogy, and biology, and in 1952 the British biophysicist Rosalind Franklin used it to analyze the DNA molecule. Her images of it were called "the most beautiful X-ray photographs of any substance ever taken."[61]

The Cavendish laboratory, which had been steered by Lawrence Bragg towards the study of biology, was well placed to exploit these results, and in 1953 James D. Watson and Francis Crick proposed their famous double-helix model of DNA.[62] It consisted of two extremely long strings made up of four bases, denoted A, C, G, and T. The strings were complementary, so a C on one always bonded to a G on the other while an A always bonded to a T.

Scientists soon pieced together what Crick called the "central dogma" which explained how DNA acted as a self-replicating program for life. Genes are segments of DNA that code for particular proteins. As with silver-based photography, production is done in two steps. First, the DNA is transcribed by the cell's machinery to make a molecule of RNA, similar to a photographic negative, and then the RNA is transcribed to make a protein molecule.

The central dogma, coupled with Neo-Darwinism, was therefore a kind of Theory of Everything for life, in which the role of stable and eternal atoms—which are governed by strict laws and acted on only by chance mutations—was played by genes. Life progressed through a series of random perturbations, rather like an ensemble forecaster taking random potshots in the hope of getting it right. As biologist Jacques Monod wrote in a 1972 book with the Democritus-inspired title *Chance and Necessity*, "Chance alone is at the source of every innovation, of all creation in the biosphere. Pure chance, only chance, absolute but blind liberty is at the root of the prodigious edifice that is evolution. . . . It today is the sole conceivable hypothesis, the only one that squares with observed and tested fact."[63]

At the time, scientists still didn't know much about what genes actually looked like because the DNA molecule was horrendously complex. In 1984, a summit was organized to discuss methods for detecting genetic mutations in descendants of the survivors of the Hiroshima and Nagasaki atomic bombs. It was proposed at the meeting that it should be possible to sequence the human genome—in other words, to determine the exact sequence of bases A, C, G, and T in a person's DNA—and use it to understand the effects of mutations. The U.S. Department of Energy was looking for ways to diversify into other "big science" projects apart from particle accelerators and thus the Human Genome Project was born.

It took a while for the technology to catch up, but the scientific effort to sequence the genome eventually morphed into a race between a publicly funded consortium and the private Celera Corporation, headed by Craig Venter, who sequenced the DNA from his own body. In 2001, after spending about $3 billion, they decided to call it a draw and cross the finishing line together. The cost of sequencing DNA has since dropped to under a thousand dollars per person.

The Human Genome Project was explicitly modeled after accelerator science. In theory, our DNA encodes the instructions for every cell in our body—just as the Standard Model tells us everything we need to know about the material universe. The sequencing technique involved dismembering the genome into small pieces (as in the Greek myth of Dionysus), figuring out the genetic code of each segment, reconstituting it all using supercomputers, and then discerning the separate genes which code for particular proteins. In aesthetic terms, it had all the qualities of good, hard, atomistic science—and at its heart was one of the most beautifully geometric molecules ever discovered.

When it was first announced, the success of the genome project was heralded as a major advance, promising what British Prime Minister Tony Blair called "a revolution in medical science whose implications far surpass even the discovery of antibiotics."[64] But while the ensuing flood of data from the genome and proteome (the network of proteins produced by that genome) certainly led to radical changes in biology, the revolution was a little different than anticipated. As discussed in the last chapter, perhaps the most interesting development was a new aesthetic based on networks and systems rather than reductionism, as well as a new understanding and appreciation of complexity, both of which will certainly have implications for health care. However, the promised medical miracles and "magic bullet" cures have been slow to arrive.

Personalized genetic tests are now available that screen thousands of genes for versions that influence disease, but as the *Guardian* reported, studies have shown that they "are inaccurate and offer little, if any, benefit to consumers."[65] Drug discovery based on reductionist methods, such as targeting single genes thought to be associated with particular health conditions, has proved in most cases to be highly expensive and unreliable. The cost of developing a successful drug has actually soared to over a billion dollars.[66]

Predictive mathematical models based on genetic or other biological information have attracted the attentions of many former physicists, engineers, and computer scientists, but have proved even trickier to produce than the Standard Model. Again, some dark matter or quintessence appears to be controlling our destiny and has proved surprisingly resistant to detection.

Box 8.2

The Tinkertoy Molecule

Watson and Crick arrived at their spiral configuration for the DNA molecule by playing around with physical models "superficially resembling the toys of preschool children."[67] They were guided by Franklin's X-ray images, their chemical and physical intuition, and also by a sense of aesthetics, or what Watson called "the belief that the truth, once found, would be simple as well as pretty." When they showed their model to Rosalind Franklin, she agreed (in Watson's words) that "the structure was too pretty not to be true."

The first artist to represent this pretty molecule in his work, just a few years after its discovery, was Salvador Dalí. His paintings on the subject included *Butterfly Landscape (The Great Masturbator in a Surrealist Landscape with DNA)* (1957–58) and *Galacidalacidesoxyribonucleicacid* (1963), and they paid homage to what he called "the only structure linking man to God."[68] However, DNA only really started to replicate in the minds of the general public after the release in 1968 of Watson's bestselling book *The Double Helix*, which described the race to discover its structure.

As Suzanne Anker and Dorothy Nelkin observe in *The Molecular Gaze: Art in the Genetic Age*, the double helix would soon become "the 20th century's iconic molecule." Its spiral motif has appeared in sculpture, architecture, and "in unexpected popular venues, such as a playground in South Beach in Staten Island, New York; sidewalk chalk figures; origami models; Lego constructions; jewellery; perfume bottles; body tattoos; and a variety of other ornaments bordering on kitsch."[69]

As a famous image from twentieth-century science, the model of the DNA molecule is rivaled only by the old solar-system model of the atom, the Trinity nuclear test, and the 1969 Earthrise photograph. Quantum physics and relativity have no similarly iconic image—unless it's the famous photograph by Arthur Sasse of Einstein sticking his tongue out.

DNA has also become part of the language: we talk about company DNA or organizational DNA. We could even say that the molecule—the longest list of opposites in the world—has become part of our scientific and cultural DNA.

To many people, including some scientists, DNA has become almost synonymous with life. In his 1976 book *The Selfish Gene*, the zoologist Richard Dawkins popularized the notion that we are just the temporary hosts of our quasi-immortal genes, which modify our behavior to maximize their chances of survival.[70] Physicist David Deutsch wrote, "An organism is the sort of thing—such as an animal, plant or microbe—which in everyday terms we usually think of as being alive. But . . . 'alive' is at best a courtesy title when applied to the parts of an organism other than its DNA."[71]

While models highlight the molecule's beautiful spiral shape, when found in nature the molecule is hidden away inside tightly twisted, compact bundles which are protected by layers of proteins. Access is tightly controlled. The molecule is also closely monitored for any damage should it come into contact with a passing cosmic ray or toxic chemical.

And while DNA is famed for its neat geometry, it is its "flaws" that drive evolution—for if the molecule were perfectly static and immune to change then Darwin's natural selection would have nothing new from which to select.

Virtually Human

One source of "dark matter" may be the genome itself. Just as only 4 percent of the universe seems to be detectable, only about 1.5 percent of the DNA molecule codes for proteins. The other 98.5 percent used to be called "junk DNA" on the basis that it didn't seem to do anything. However, we have since learned that it plays a vital (but still poorly understood) role in tasks such as regulating gene production.[72] So-called epigenetic factors also play a role: DNA is just a molecule which has to be read and interpreted by a host of other molecules.[73]

While some rare diseases such as cystic fibrosis are caused by a single mutated gene, most conditions are better viewed as the emergent result of a mix of multiple genetic, epigenetic, environmental, psychological, and social factors. The more we learn about human biology, the more we find that DNA is just one partner in a very complicated dance. And, as with the atmosphere, biological systems are characterized by complicated structures of positive and negative feedback loops which regulate and maintain order in the system. Thus a fault or perturbation in one part of the system may be compensated, or enhanced, in another.

The central dogma of biology, which sees life as a kind of molecular factory controlled by DNA, implies that we should in principle be able to build a Newtonian model that tracks the production and interactions of proteins within the body in the same way that weather models track the flow of air, water, and heat in the atmosphere. Indeed, there are several projects that are attempting to build a fully mechanistic model of a human being. The 2008 Tokyo declaration from the Japanese Systems Biology Institute called for "a grand challenge project to create over the next thirty years a comprehensive, molecules-based, multi-scale, computational model of the human ('the virtual human'), capable of simulating and predicting, with a reasonable degree of accuracy, the consequences of most of the perturbations that are relevant to healthcare."[74] Similarly, the European Union has its Virtual Physiological Human.[75] Oak Ridge National Laboratory, which was first established as part of the Manhattan Project, also has a project it calls the Virtual Human, although most of its efforts appear to have been diverted to a side project called Virtual Soldier.[76]

This approach conforms to the mainstream scientific aesthetic that a model should be built up from first principles—for example, genes and proteins—to simulate the entire system. If positioned on the wall next to the climate model pastiche, it will blend in perfectly. However, while such models are useful for understanding how the parts of the system relate to one another, as working models they tend to be complicated and unstable. They are therefore not very useful for making predictions and have so far proved hard to commercialize.[77]

Perhaps because biologists have traditionally been less influenced by physics than some other areas of science, they have also been open to a range of new modeling tools. As discussed further in the last chapter, these include methods from new areas of applied mathematics such as complexity theory and network theory. Sophisticated techniques have been developed to spot statistical patterns in biological data, and these can be used for example to predict which patients will respond well to particular treatments.

One area where genetic testing and modeling holds great promise is in the treatment of cancer, but here the genome of the cancer cell matters more than the genome of the patient. Cancer is in a sense a genetic disease because cancer cells carry genetic mutations that allow them to escape the cell's usual regulatory controls and reproduce without limit, like a nuclear fission reaction in slow motion. Different types of cancers are characterized by particular mutations with selective responses to drugs. By performing a genetic analysis on tumors, it should be possible to tailor the drug regime for individual patients.[78]

An unfortunate consequence of our mechanistic approach to health care is that it implies a certain ranking to medical conditions. Diseases which have a clearly mechanistic origin appeal to both the medical profession—because they are amenable to logical treatment—and to patients, who often believe that associations with any kind of mental issue represents a loss of status. However, many conditions have a strong psychosomatic component. Indeed, one of the world's miracle drugs—which in clinical trials of new treatments is used as a hard-to-beat benchmark—is available completely free. It's called a placebo, and no one knows exactly how it works, but it doesn't have much to do with genes.[79]

The first public medical trial of the placebo effect was carried out in 1784, at the request of King Louis XVI, by a commission that included Benjamin Franklin (then the American Ambassador to France) and the chemist Antoine Lavoisier. Their subject was the work of one Franz Anton Mesmer, who claimed to be able to endow objects with an "animal magnetism" that had curative powers.[80] After handling a treated object or substance, his clients—who included Marie Antoinette—would enter a "crisis" marked by convulsions, crying, fainting, and so on, after which they were pronounced cured. The King suspected a fraud, and indeed the commission found for example that some patients experienced a crisis after touching plain water which they were told had been "mesmerized" (the word is from Mesmer's name) while others did not react to treated water which they were told was plain. The commission concluded that any effects were due to "imagination."

While the rational approach of Franklin and Lavoisier put an end to Mesmer's career, the placebo effect lives on exactly because of the power of imagination.[81] Health care has become polarized between conventional medicine, which seeks mechanistic solutions and is mesmerized by the beauty and magnetic appeal of DNA; and alternative approaches such as homeopathy, which harness the placebo effect in anyone who lacks or can repress a basic education in science. As Ted Kaptchuk of Harvard Medical School notes, "The goal is to understand placebos so that they may be used intelligently. . . . We need to stop pretending it's all about molecular biology. Serious illnesses are affected by aesthetics, by art, and by the moral questions that are negotiated between practitioners and patients."[82] The greatest source of dark matter for health care may be our own brains.

Fixed-Point Economics

While biologists have proved open to new modeling techniques, when it comes to simulating the larger biological system known as human society, economists have taken an approach that seems positively Aristotelian in its aesthetics. Just as Aristotle's Theory of Everything was really a theory of stability, so mainstream neoclassical economics is based on the idea that the complex, dynamic, evolving, and crash-prone economy is best modeled by assuming it yearns towards a stable resting place.

Neoclassical economics was first developed in the late nineteenth century and was explicitly modeled after Newton's "rational mechanics." As discussed in Chapter 6, Newtonian dynamics can be expressed through the calculus of variations as an optimization problem: objects moving in a field take the path of least action. Leibniz had explained the idea by comparing God to an architect who "utilizes his location and the funds destined for the building in the most advantageous manner."[83] Reasoning along the same lines, neoclassical economists assumed that in the economy, individuals act to optimize their own utility—defined rather hazily as being whatever is pleasurable for that person—by spending their limited funds.[84] Economists could then make Newtonian calculations about how prices would be set in a market economy to arrive at what William Stanley Jevons called a "mechanics of self-interest and utility."[85]

One reason why mathematics works so well in physics is that, as far as we know, subatomic particles such as electrons and quarks are the same everywhere in the universe. As a result, a hydrogen atom on Earth is the same as one in the Sun. People, on the other hand, are different. To get around that problem, economists argued that what really counted was the behavior of the "average man." This concept was first introduced by the French sociologist Adolphe Quetelet, who saw the average man as representing "perfect harmony, alike removed from excess or defect of every kind . . . the type of all which is beautiful—of all which is good."[86]

It was certainly safe to assume that this human version of Worthington's "Auto-Splash" made rational decisions based on entirely selfish and individualistic motives. As economist Francis Edgeworth put it, "the first principle of Economics is that every agent is actuated only by self-interest."[87] Thus was born *Homo economicus*, or "rational economic man"—an idealized expression of Nietzsche's Apollonian *principium individuationis.*[88] Using this imaginary being as the atom of the economy, economists argued that in a competitive market prices would be driven to a stable equilibrium via Adam Smith's invisible hand: if a particular good were too expensive, then more suppliers would enter the market and competition would drive the price down; if prices were too low, then suppliers would go broke or leave and the price would rise.[89] The result, according to Edgeworth, would be "the maximum of pleasure" for both individuals and society as a whole.

While rational economic man is obviously a caricature, he has been a surprisingly influential one. In the 1940s, John von Neumann used him as the basis for his game theory, which studied the interactions between rational actors who are trying to optimize their own outcomes in artificial games. As von Neumann's biographer Steve Heims notes, game theory "portrays a world of people relentlessly and ruthlessly but with intelligence and calculation pursuing what each perceives to be his own interest. Other 'players' are seen as enemies, competitors or collaborators."[90] Of course, this also fits quite well with the clinical definition of a sociopath. Game theory was first developed for economics, but it is perhaps best known for its justification for the doctrine of Mutually Assured Destruction, which was discussed in Chapter 5.

Fresh from his triumph of protecting human survival from nuclear conflict, rational economic man was soon helping to win the ideological battles of the Cold War (indeed, much of the financial support for economics came from defense-related U.S. government spending). In the 1960s, the economists Kenneth Arrow and Gérard Debreu used a method popular in game theory known as Brouwer's fixed-point theorem to prove that, under certain conditions, free markets lead to an optimal "fixed point" for the economy in which prices are set at their correct levels and nothing can be changed without making at least one person worse off (a condition known as Pareto optimality).[91] This result, which reminds one of Alberti's fifteenth-century definition of beauty as a harmony of parts in which any change is for the worse, was soon being claimed as proof that capitalism was superior to communism. But to accomplish this feat, the powers of rational economic man had to be extended to include infinite computational power and the ability to devise plans for every future eventuality. The clarity and beauty of his mind were indeed approaching that of Apollo.

The Arrow-Debreu model has been called the crown jewel of neoclassical economics, and it inspired the development of General Equilibrium Models which are still relied on by policymakers today. Like weather models, these split the model up into manageable units—in this case, sectors by region or industry—which are assumed to optimize their own utility. Economists can then study how, for example, a change in government policy will affect the economy's equilibrium. Unfortunately, numerous studies have shown that their predictive accuracy is not much better than random guessing.[92] (A graphic example of the inability of economists to predict the future was revealed in the minutes of the U.S. Federal Reserve from late 2006, when the housing bubble was already starting to pop. No one suggested the possibility of even a mild recession, and the consensus seemed to be that the economy was "a lot like a tennis racquet with a gigantic sweet spot."[93])

These drawbacks didn't mean the model was wrong, though. Perhaps the greatest triumph of rational economic man was to accomplish something that no economist had previously been able to do—which was to explain why economic predictions were so poor.

Perfect Model II

Eugene Fama's 1965 efficient market hypothesis saw the market as consisting of "large numbers of rational profit-maximizers" who have access to perfect information.[94] It claimed that market forces, acting like a perfectly efficient machine, would drive the price of any security to its "intrinsic value." Any changes were small and random, or were driven by unpredictable news. It was therefore impossible to beat the market or predict future changes, as all the information was already priced in.

The efficient market theory has been one of the most influential ideas in economics. *The Economist* described it as "a building block for other theories on subjects from portfolio selection to option pricing."[95] Warren Buffett once ironically compared it to "holy scripture."[96] As discussed below, however, it enjoys no empirical support. Its popularity amongst economic modelers can be explained by the fact that it did exactly what the butterfly effect did for weather modelers: it gave an excuse for prediction error and helped to preserve the idea that models are not flawed.[97]

As with meteorology, the lack of predictive accuracy did not imply that economic forecasters were out of a job—only that they should shift to making probabilistic forecasts. Since the market was assumed to be at or near equilibrium, and since price changes were random, price fluctuations could be modeled as random perturbations to a steady state. The motion of the market was therefore similar to the so-called Brownian motion of a small dust particle as it is buffeted around by collisions with individual atoms. This can be modeled using the statistical technique known as the normal distribution, or bell curve. One of Einstein's annus mirabilis papers used the technique to estimate the size of an atom (its use in finance actually came first, in a PhD thesis by Louis Bachelier in 1900, but it didn't catch on at the time[98]). So, rational economic man had literally become the atom of the economy and human variation was normalized away. As noted by the author and derivatives trader Pablo Triana, "all the Nobels awarded to financial economics are heavily grounded on the Normal assumption; remove such tenet, and the prized theories crumble and crash."[99] Even today, risk models used by institutions like banks or companies continue to be based on this mathematics, with at best minor modifications.

Rational economic man reached his highest state of perfection with the rational expectations theory of Robert Lucas, who was Fama's colleague at the University of Chicago. This assumed not only that market participants were rational but also that they had a perfect model of the economy in their head, in the sense that they didn't make systematic errors. As with the efficient market hypothesis, the theory assumed that markets were at a static equilibrium; if prices were too high or too low that would imply that people were not being rational.

The idea of rational behavior was also given a credibility boost in the 1970s by Richard Dawkins, who provided a link between genetics and natural selection. As he wrote in *The Selfish Gene*, "If you look at the way natural selection works, it seems to follow that anything that has evolved by natural selection should be selfish."[100] We are rational, utility-maximizing machines because our genes are. An implication of this was that economic success reflected superior genes. The psychologist Oliver James notes that the book's rise to bestseller status in the 1980s coincided with Reagonomics as well as Thatcher's "no such thing as society" ethos.[101]

It is ironic that an economic theory gained credibility from natural selection, since the entire idea behind natural selection was inspired by the economist Thomas Robert Malthus, who argued that population growth inevitably results in food shortages and a brutal competition for survival.[102] The relationship between economic survival and natural selection was made explicit by social Darwinists such as Herbert Spencer, who argued that charity would only prevent society from "excreting its unhealthy, imbecile, slow, vacillating, faithless members."[103]

Today, mainstream economic theory continues to rely on the related and interdependent concepts of rationality, stability, and efficiency. Of course, no one thinks that markets are perfectly stable or that people are perfectly rational, and even orthodox economists are quick to distance themselves from such simplified caricatures. Much work has been done to explore deviations from perfection. But these core assumptions underlie the equilibrium models used by policymakers and the risk models used by banks. They are held as a kind of ideal to which real markets can aspire. They provided an excuse for the general trend towards deregulation in the financial sector that occurred up until the late 2000s. They have also affected human behavior: the ideal of our age—at least in business schools—is that of the rational, stable, efficient, corporate hero who puts money above all else. When a trader told the BBC during the 2011 European sovereign debt crisis that "I have a confession, which is I go to bed every night, I dream of another recession" because of the money-making opportunities it afforded him, he could be viewed as speaking with the clear logic of his selfish genes.[104] We are willing to tolerate antisocial behavior from those imaginary legal entities known as "corporate personhoods"—the closest thing we have to rational economic man—because that is the logic of the economy.[105] We have internalized values such as independence and materialism, when happiness has more to do with connectedness and community. No wonder reported happiness levels have been in decline in industrialized countries since the 1960s.[106] For a theory that was supposed to achieve the "maximum of pleasure," neoclassical economics has failed to deliver.

Like the ancient Greeks, we model the world as if it is a stable, ordered, rational, beautiful machine, ruled by what the Romantic artist and poet William Blake called "fearful symmetry."[107] The only difference is that the circle-based models used by the Greeks could predict things like solar eclipses—they could give some warning that the lights were going to go out. Our models lack that kind of empirical validity.

The reason that Aristotle's philosophy continued to dominate in the Middle Ages, according to the author James Hannam, was that it offered "a complete system of reality," so that "rejecting any significant chunk of it would cause the whole edifice to collapse."[108] Tottering on its pillars of rationality, stability, and efficiency, the same could be said of mainstream economics today—and it's about to collapse.

Dark Energy

So-called heterodox economists, as well as people like mathematicians and physicists, have long argued that the assumptions of neoclassical economics are absurd.[109] One of the first to mock the idea of rational economic man was Thorstein Veblen, who in 1899 compared him in comic-opera language to "a lightning calculator of pleasures and pains who oscillates like a homogeneous globule of desire of happiness under the impulse of stimuli that shift him about the area, but leave him intact."[110] More recently, behavioral economists have explored the role that emotion plays in decision-making.[111] As an example, one 2010 study showed that when ordering desserts at a restaurant, we will pay on average about 50 percent more if they are offered on a dessert cart rather than a menu.[112] The reason is that when the food is there in front of us, a completely different part of our personality is engaged—for a moment we become rather like a five-year-old child who wants immediate gratification (a tendency which is fine in restaurants, but less good for financial planning). Indeed, neuroscientists have shown that the way options are presented affects which part of our brain we use to handle them (though when it comes to debunking rational economic man, this is like using an MRI machine to crack a nut).[113]

The other assumptions of neoclassical economics are equally unrealistic, and like the butterfly effect they seem to resemble a kind of magical thinking about the system more than they resemble science. The idea that markets are efficient because they are unpredictable makes no sense; snowstorms are unpredictable but no one says they are efficient. As for market stability, which the efficient market theory is based on, that is a belief that would please Aristotle, but few hedge funds.[114] A number of studies show that price fluctuations are not random perturbations around a steady state, but follow a so-called power-law distribution. As will be discussed in Chapter 10, this is a signature of many complex systems operating at a state which is best described as *far* from equilibrium, at the border between order and chaos.[115] Rather than being static, the economy is dynamic, creative, evolving, and subject to sudden changes. It is also driven by numerous factors, which together constitute a kind of dark energy that completely eludes economic models. Five of the main components are love, power, justice, money—and everything else.

Love. The anthropologist Gregory Bateson once said, "Of all our human inventions, economic man is by far the dullest." His daughter Mary Catherine Bateson later commented, "We can be grateful that in this case no one has cleaned up his gender biased language because the concept is not and never was a gender neutral one. The dangerous idea that lies behind 'economic man' is the idea that anyone can be entirely rational or entirely self-interested. One of the corollaries, generally unspoken in economics texts, was that such clarity could not be expected of women who were liable to be distracted by such things as emotions or concern for others."[116] Tasks such as looking after the young, old, and sick—which are usually carried out by women—are often paid poorly or not at all, but are still vital to the health of society. Much of our sense of wellbeing comes from social relationships, but this is not reflected in economic metrics of utility. Selfish gene theory, which helped to rationalize rational economic man, ignores the fact that selection acts not just on the level of genes or individuals, but also on groups and social structures. As the Santa Fe economist Samuel Bowles has shown, this type of selection leads to "the proliferation of group-beneficial behaviors" that are "quite costly to the individual altruist."[117] Like subatomic particles, people are entangled. None of this appeals to the individualistic, male aesthetic of neoclassical economists, who economist Deirdre McCloskey compares to a motorcycle gang, "strutting about the camp with clattering matrices and rigorously fixed points, sheathed in leather, repelling affection. They are not going to like being told that they should become more feminine."[118]

Power. Neoclassical economics treats individuals and firms as more-or-less identical atoms. If people behave like rational economic man, that means they will all make the same rational decisions and have approximately the same advantages. This assumption ignores the complex power relationships in the economy. As Norbert Häring and Niall Douglas wrote in *Economists and the Powerful*, there is "the power to abuse informational advantage, the power to give or withhold credit, the power to charge customers more than it costs you to produce, and the power to change the institutional setting to your advantage. There is the power of the corporate elite to manipulate their own pay and to cook the books, the power of rating agencies to issue self-fulfilling prophecies, the power of governments to manipulate the yardsticks that voters are offered to judge their economic policies. All these types of power, which were important in bringing about the global financial crisis, are defined away by standard assumptions of most mainstream economic models. These models feature perfect competition, efficient financial markets, full information and eternal equilibrium."[119] This stance justifies the existing power structure by implying that those in power are there because they have out-competed everyone else in a fair fight and deserve their reward.

Justice. By posing as an objective science, neoclassical economics seemed to remove the need to take ethics into account. The invisible hand of the marketplace was assumed to deliver a reasonably just and equitable distribution of wealth. As a result, writes economist Tomáš Sedláček in his book *Economics of Good and Evil*, "ethics has disappeared from mainstream economic thought."[120] However, excessive concentration of wealth in the hands of a few violates our basic sense of fairness, with destabilizing effects on society and the economy.

Money. Following the award of a Nobel Prize in 1921 for his work with Rutherford on radiation, Frederick Soddy switched his attention from physics to economics. One of the few people to correctly predict that nuclear weapons would soon be developed, Soddy realized that the world was not a safe place for them, in large part because of inadequate economic theories. At the root of the problem, he believed, was a confusion between real wealth and what he called "virtual wealth," namely money and its derivatives.[121] Real wealth (like uranium) is subject to decay, but money obeys abstract mathematical laws and can grow without limit. Because it is a form of debt, it represents a negative quantity, which does not exist in nature (unless you count antimatter). Under fractional reserve banking, banks effectively create money out of thin air (big-bang-style) by issuing debt that is only partly backed by real assets, so virtual wealth soon exceeds real wealth. This is highly unstable, because when a crisis occurs everyone wants to cash in at the same time. The result can be social instability—or war.

To reduce the size of the virtual economy, Soddy proposed a 100 percent reserve requirement for banks. A first step in the twenty-first century would be to model it properly—for perhaps the weirdest feature of the models used in mainstream economics is that money doesn't play much of a role in them. The Arrow-Debreu model didn't include the financial system because it was assumed that capital markets worked perfectly and always cleared. Banks only acted as intermediaries, and so their role could be ignored. In a perfect, Platonic world, no one needs to dirty their hands with banknotes. As a result, most General Equilibrium Models—of the type used, for example, by the Bank of England—don't include banks or hedge funds either.[122] This made it hard for the models to predict the recent banking crisis. (See Figure 10.3 for another approach.)

Everything else. Finally, mainstream economic theory fails to pick up an even larger source of dark matter and dark energy, which is everything outside the monetary system, including the Earth. The environmental crisis, which includes global warming, is a byproduct of our economic system—but because models typically ignore or downplay things like resource depletion, environmental damage, or the beauty and diversity of nature, it barely even registers. From an environmental perspective, the image of the economy as rational and efficient therefore seems particularly delusional. The "heterodox" field of ecological economics is an attempt to correct these omissions.[123]

As with physics, economics has been shaped by the quest for abstract beauty. Milton Friedman, whose rational free-market philosophy was enormously influential on many politicians including Thatcher and Reagan, was inspired to study mathematics by a high school geometry teacher who connected Keats's "beauty is truth, truth beauty" with the Pythagorean theorem.[124] Again, as with physics, this mathematical elegance was achieved by imposing symmetry on the system. Rationality is a symmetry because rational people with the same preferences will make the same decision given the same information. Stability is symmetry in time—if markets are in equilibrium, then the future looks like the past. And if market participants have similar power and other characteristics, then that means transactions are symmetric.

Of course, any model makes simplifications—and, as Friedman pointed out, if the aim is to make accurate predictions then the realism of the assumptions is of secondary importance.[125] But given that the model cannot make predictions, support for the assumptions comes only from their status as cultural totems. Noether showed that symmetry leads to conservation principles. In economics, what has been conserved is a world view.

In a 2009 article entitled "How Did Economists Get It So Wrong?" Paul Krugman answered his own question by saying that they "mistook beauty, clad in impressive-looking mathematics, for truth."[126] The chief economist at the International Monetary Fund, Olivier Blanchard, told a Washington conference in 2011 that the mainstream "had converged on a beautiful construction" to explain the economy, but now had to admit that "beauty is not synonymous with truth."[127] While the fascination with elegant ideas and equations has been mostly harmless in areas such as weather forecasting or genetics, in economics it has done direct damage to people's lives, as I and others have argued elsewhere, by helping to perpetuate a system that is unfair, unstable, and unsustainable.[128]

Aesthetics alone doesn't explain why neoclassical theory has survived for so long. As with Aristotle's theory of stasis, another attraction of this beautiful model is its compatibility with the aims of the richest elites. After all, if markets aren't rational or optimal, that makes it much harder to explain why a tiny sliver of the world's population should control most of the wealth while the majority labor to pay off increasingly arduous debts.[129] Like a pair of Gucci shoes from a luxury store, or a diamond-encrusted artwork by Damien Hirst, this beauty and elegance is not for the masses.[130]

Frederick Soddy observed in 1934 that "orthodox economics has never yet been anything but the class economics of the owners of debts."[131] The close connection between Wall Street firms and university economics departments, documented in the 2010 film *Inside Job*, would not have surprised him. The model has a comforting appearance of solidity and fits in well with the other pieces, but it has turned out to be a commissioned fake.

Mainstream economics is long overdue for a revolution—and, as will be discussed in the final chapter, it appears to finally be underway. Ideas from areas of applied mathematics such as network theory and complexity, coupled with insights from psychologists, ecologists, and heterodox economists, are helping to create a new economics whose inspiration—and aesthetics—is drawn from nature rather than machines. New galleries are opening up in the cheap part of town and the art doesn't look the same.

World View

As mentioned above, the main aim and test of reductionist science is to make predictions. But when mathematical models, which were first developed in physics, have been applied to things that we really want to predict—such as the climate, our health, or the economy—the results have been consistently disappointing. In each case, the approach has been to break the system down into separate components, figure out the equations which supposedly govern them, and build a large and impressive mathematical model. A degree of randomness is incorporated to explain variability and change. This meta-model conforms to the mainstream scientific aesthetic of hard equations and logical laws, but it has little predictive accuracy. The reason, it seems, is related to a kind of category error: we model living systems as if they were machines. Our ensemble simulations are like armies of unstable robots that try to capture the intricacies of natural behavior but soon wander off in the wrong direction.

Rather than abandoning the approach, scientists obfuscate the results in scientific-looking but ultimately misleading probabilistic uncertainties. The result is climate models that the public is justifiably skeptical about, virtual humans that are always ten or twenty years away from going live, and economic risk models that actually make the system riskier by creating a feeling of false confidence. At a time when humanity faces numerous environmental and economic challenges, our mathematical models are proving not to be of much help.

But what has remained surprisingly safe throughout this process, curled up like DNA in its protective sheath, is the idea of the perfect model.

As Steven Weinberg wrote, the ultimate product of physics is not just theories about subatomic particles, or a way of extending the power of our senses through telescopes and colliders, but "a world view."[132] Einstein noted, "Whether you can observe a thing or not depends on the theory which you use. It is the theory which decides what can be observed."[133] Even deeper than theories, though, is aesthetics. The way that we see the world has been shaped by the traditional scientific aesthetic. We look for reductionist theories that can break a system down into its components. We seek elegant equations that can be rigorously proven using mathematics. We aim for a unified, consistent theory. We celebrate symmetry, clarity, and formal beauty. Predictive accuracy is supposed to be the test of reductionist science—but when accurate predictions prove impossible, aesthetics wins out every time. What we mean by "good science" is strongly related to what we (i.e., scientists) mean by beauty.

Lee Smolin observed that "the extent to which people can invent imaginary worlds when science gets decoupled from experiment is quite extraordinary. They follow a certain aesthetic of mathematical elegance out there as far as it takes them."[134] And it seems that even observations by themselves are not enough to dislodge an aesthetically appealing theory—we can always invent excuses. What is required, at minimum, is another model that makes better predictions. But if the system is fundamentally unpredictable, then no such model exists. The science machine is left stuck, its wheels spinning. Like medieval theologians debating the finer points of Aristotelian physics, or theatre workers arguing over the staging of an opera without worrying that the opera itself is bad, scientists argue and contest over details, but nothing changes. In order to progress, science needs a new impulse from the outside—and if it doesn't come from experiment, perhaps it can come from ideas about aesthetics. As artist David Hockney said, "You can't be tired of nature. It is just our way of looking at it that we are tired of. So get a new way of looking at it."[135]

Aesthetics is not just about beauty; it is also about a way of perceiving. Our models are like a brilliant torch with a deep and penetrating beam. Like the Standard Model, they only seem to be able to pick out a small fraction of what is really going on. We are surrounded by dark forces that we can neither predict nor control. Models have proved most successful when used as specialized tools, adapted for particular niches, rather than general one-size-fits-all solutions. "The multiplicity of models," noted mathematical ecologist Richard Levins in 1966, "is imposed by the contradictory demands of a complex, heterogeneous nature and a mind that can only cope with few variables at a time; by the contradictory desiderata of generality, realism, and precision; by the need to understand and also to control; even by the opposing aesthetic standards which emphasize the stark simplicity and power of a general theorem as against the richness and the diversity of living nature. These conflicts are irreconcilable."[136]

Woody Allen asked, "Can we actually 'know' the universe? My God, it's hard enough finding your way around in Chinatown."[137] As seen in the next chapter, however, the fact that we cannot successfully model events on our planet has not imbued all scientists with a deep sense of humility. Instead, we have extended our models to include the birth not just of our own universe, but of others as well.

9

The Virtual Universe

We have glimpsed how He designed the universe;
now we imagine that we, too, can design universes.

Anthony Zee, *Fearful Symmetry*

But this is an old and everlasting story: what happened in old times with the Stoics still happens today, as soon as ever a philosophy begins to believe in itself. It always creates the world in its own image; it cannot do otherwise.

Friedrich Nietzsche, *Beyond Good and Evil*

There is a theory which states that if ever anyone discovers exactly what the Universe is for and why it is here, it will instantly disappear and be replaced by something even more bizarre and inexplicable. There is another theory which states that this has already happened.

Douglas Adams, *The Hitchhiker's Guide to the Galaxy*

The division between physics, which evolved with engineering, and biology, which evolved with medicine and agriculture, is as old as science. However, scientists are discovering that physical and biological systems have much in common—the line that separates animate from inanimate is losing some of its definition. Indeed, the most remarkable feature of the universe is the way that its forces are balanced so as to support complexity and life. Like a living organism, all the parts of the universe operate as a cohesive whole. This chapter argues that we have been led astray by an emphasis on machine aesthetics.

In February 1738, at the fair of Saint-Germain in Paris, Jacques Vaucanson presented his version of a virtual human to the public. The exhibit was called the *Automaton Flute Player* and it consisted of a life-size android figure, made of wood and seated on a rock, who could play twelve different tunes on a transverse flute. A complicated mechanism involving nine separate bellows piped air into the mouth. The flow was modulated by a movable metal tongue, which along with the lips and fingers was controlled by a system of pulleys and levers. The machine was powered by a weight-driven mechanism hidden in the base. The wooden fingers were coated with leather to give smoother contact. All of the motions were perfectly harmonized to give the impression of a real human player.

The effect was evidently impressive, since Vaucanson could charge three livres for it—this was a week's wages for many workers. The Mercure de France noted that "everything, including crescendi, diminuendi, and even sustained notes, is executed with the most perfect good taste."[1] He presented his machine to the French Academy of Science, who praised him for the "clarity and precision" of his work. Voltaire's 1738 poem *Discourse in Verse on Man* compared him to Prometheus:

> With a hand on which all praise falls sterile,
> Vaucanson the bold, Prometheus' rival,
> Took, while imitating nature's projects,
> Heaven's fire to animate cold objects.[2]

In the next year, Vaucanson produced two more automata. The first was *The Tambourine Player*, which accompanied its own flute-playing with the tambourine. The second was Vaucanson's most famous creation: *The Digesting Duck*.

The duck was made of gold-plated copper and had over a thousand moving parts. It could drink water, quack, raise itself on its legs, and eat seed from the hands of spectators with a realistic gulping motion. But this machine went beyond his other automata because it represented not just the duck's external but also its internal functions. As he explained, in perhaps a little more detail than necessary, "the food is digested in the stomach as it is in real animals. . . . The material digested in the stomach passes through tubes, as it does through the entrails in the animal, to the anus, where there is a sphincter to allow its release."[3] If you put food in, a short time later a kind of green gruel would be discharged from the bottom.

The Digesting Duck was a huge hit with the public and Vaucanson grew rich by taking it on tour. By 1741, though, he had grown tired of his creation and sold it to three Lyonnais businessmen. Louis XV had been impressed by the duck, and he gave Vaucanson a new and more prestigious job: inspector of silk manufacture. The silk industry in France was falling behind its competitors in England and Scotland, and it needed modernization. Vaucanson would go on to invent the first fully automated loom. This time his invention received a less rapturous response—he was pelted with stones by weavers who believed the machine threatened their jobs.

The famous duck, meanwhile, was passed around between impresarios, collectors, and pawnbrokers for 150 years until it is thought to have been destroyed in a fire in 1879. One of the people to encounter it was the magician Jean-Eugène Robert-Houdin (who Houdini named himself after). On examining the duck at the Paris Exposition Universelle in 1844, the magician noted that the defecation was actually "a piece of artifice I would happily have incorporated in a conjuring trick."[4] The duck turds were green-dyed breadcrumbs formed into pellets, which were kept in a separate compartment and released a set time after food was consumed.

Vaucanson's creation raised questions for his audience which continue to haunt us today. What is it that separates the living from the machine? Can man create life? And if something looks like a duck, walks like a duck, and quacks like a duck, is it really a duck?

Machine Aesthetics

In many respects, predictive mathematical models of the sort discussed in the previous chapter resemble Vaucanson's virtual duck. The atmosphere, the human body, and the economy are all complex organic systems which are either alive or the product of life (the atmosphere is created by the respiration of plants and animals). Our models are superficially beautiful, with their gold-plated equations and thousands of moving parts; but their workings don't stand up to close analysis and their output—their green-dyed breadcrumbs—isn't the real thing. So what about the universe itself, the source of all life? Can we model that as a machine?

The aim of physics is not just to simulate the appearance of visible matter. The Standard Model has done that perfectly well for some decades now, and everyone applauds its ability to quack, or quark, on demand. But we want the unique set of equations—recognizable by its clarity and beauty—that makes the universe *inevitable*. The efforts to unify the four forces, and to extrapolate the known laws of physics to energy levels that can never be tested, are all based on this effort to mechanize the cosmos—and, as argued here, they reflect aesthetic choices. The psychiatrist R.D. Laing wrote that "Galileo's program offers us a dead world: Out go sight, sound, taste, touch, and smell, and along with them have since gone esthetic and ethical sensibility, values, quality, soul, consciousness, spirit."[5] But our aesthetic sensibility has not completely atrophied—it has just been taken over by a machine.

In the worlds of art and design, machine aesthetics reached the peak of its popularity in the 1920s and 1930s. In his 1934 book *Technics and Civilization*, the historian and philosopher Lewis Mumford's list of "machine canons" included the "elegance of a mathematical equation," the "inevitability of a series of physical inter-relations," and "the tight logic of the whole."[6] Susan Fillin-Yeh described the 1920s machine aesthetic as "functional in design, with . . . a Euclidean, geometric perfection, and neatness."[7] W.H. Mayall added "simplicity of outward form" and "the need for unity of overall form."[8] Of course, they were talking about art objects or designed products like cars and kitchen blenders, but the same set of aesthetic concepts appears quite general. When it comes to our scientific ideas, we are all driving around in vintage Chrysler Airflows with extra streamlined features such as supersymmetry to give the appearance of functionality. Our highest scientific temple—the Large Hadron Collider—is also the world's biggest machine.

The machine aesthetic has been particularly influential in the area of genetics. At a 1991 Royal Institution Christmas Lecture, given to an audience consisting largely of children, Richard Dawkins explained, "We are machines built by DNA whose purpose is to make more copies of the same DNA. . . . That is EXACTLY what we are for. We are machines for propagating DNA, and the propagation of DNA is a self-sustaining process. It is every living object's sole reason for living."[9] The idea that we are utility-maximizing machines also entered economics through its concept of rational economic man. Our sole reason for living is to earn money because that's what our gene machines—our coils of DNA—are telling us to do.

As Laing points out, though, our machine aesthetic comes at a cost. One of the differences between living things and machines is that we tend to bond better with the former. Our vision of the universe—and of ourselves—as nothing but a deterministic machine is therefore cold, distancing, and rather depressing. According to this picture, noted Alfred North Whitehead, "Nature is a dull affair, soundless, scentless, colourless; merely the hurrying of material, endless and meaningless."[10] Steven Weinberg wrote, "The more the universe seems comprehensible, the more it also seems pointless."[11] And Stephen Hawking contended that "the human race is just a chemical scum on a moderate-sized planet."[12]

Behind all the enthrallment at the formal beauty of the universe, and past the default optimism of scientists with its belief in eternal technological progress, there lies a rather bleak world view that has had far-reaching effects. The search for beauty and inevitability in equations makes us in comparison seem like ugly accidents—mere random mutations. As political scientist Robert Wesson wrote, "There is something of self-hate in the materialist approach. It depreciates the life of the mind and works of imagination and character. It demeans the richness and wonder of nature. It seems to make unnecessary further thinking about the mysteries of existence, of life and the universe."[13] Perhaps this explains why our economic system seems so intent on turning the world into a productivity-maximizing machine, with humans as cogs. Einstein, who seems to have experienced a change in heart about science after the war, lamented that "the horrifying deterioration in the ethical life of people today stems from the mechanization and dehumanization of our lives—a disastrous by-product of the scientific mentality.[14]

So why is there this anxiety to make everything automatic, to get rid of any hint of purpose or intent? Is it just another skirmish in the long battle between religion and science, or is there also something else at work?

Box 9.1

Machines for Living

As the philosopher Mark C. Taylor notes in his book *The Moment of Complexity: Emerging Network Culture*, the modernist machine aesthetic is based on the idea of the grid. Grids are straight, static, simple, symmetric, uniform, and functional. They only have a single scale, because each square of the grid is the same size. In classical physics, atoms exist on a grid—especially when aligned in a crystal. Rational economic man lives on a grid. So do many of the rest of us, since cities are often planned around a grid. Our working days are also divided into a one-dimensional grid of time points. Grids don't have to be dull—painters from Dürer to Mondrian to Chuck Close have exploited the grid's ability to provide a backbone of order while allowing for variety.

Living systems, in contrast, tend to be based on networks. These are nonlinear, dynamic, asymmetric, scale-free (in the sense that there is no typical size or level of importance), and organic. The World Wide Web is organized as a network, as are ecosystems or the financial network. As was mentioned in Chapter 7, the theory of loop quantum gravity views space as structured like a network.

In architecture, the champion of grids—at least for part of his career—was Le Corbusier, whose imposingly beautiful modernist buildings were intended to be what he called "machines for living." As he wrote in 1929, "Machinery is the result of geometry. The age in which we live is therefore essentially a geometrical one; all its ideas are so orientated in the direction of geometry."[15] The modern city "lives by the straight line, inevitably; for the construction of buildings, sewers, and tunnels demands the straight line; it is the proper thing for the heart of a city. The curve is ruinous, difficult and dangerous; it is a paralyzing thing." He argued that the unplanned continental city was a kind of accidental monstrosity that should be leveled and replaced with something based on a formal geometric layout (though some of his later work, such as the chapel of Notre Dame du Haut in Ronchamp, France, appears more inspired by organic form).[16]

Modernist architecture yielded to the movement known as postmodernism, which mixed and matched elements from various schools in a somewhat haphazard and ironic way; and this in turn gave way to a generation of architects such as Rem Koolhaas, Zaha Hadid, and Frank Gehry, whose work is decidedly nonlinear and off-grid. As Taylor writes, Gehry's Guggenheim Museum Bilbao is based on a "complex horizontal network of dispersed yet interrelated forms. . . . When Euclidean geometries are fractured, surfaces left unfinished, and forms rendered incomplete, structures are opened in ways that allow complexity to emerge in surprising ways. Having found the mechanistic logic of modernism inadequate, Gehry seeks an alternative logic that approximates the logic of networking."[17] While his designs rely on advanced machines and computer engineering, he "subtly folds the mechanical logic of industrialism into his work in ways that paradoxically negate and preserve its traces."

This might seem to have little to do with science, but Gehry also spends a few weeks a year working with Princeton University's department of microbiology to see if his creative methods can loosen up what he sees as a rigid approach to cancer research. As he told the *Observer*, "There has been so much funding and so much science in big institutions in 30 years, but it hasn't moved the needle."[18]

Perhaps the greatest recent demonstration of the power of networks has been the role that computer networks and social media have played in political uprisings such as the Arab Spring and Occupy movements. As J.G. Ballard noted, great cultural shifts are largely aesthetic: today the shift is networks taking over the grid.

Figure 9.1 Albrecht Dürer's perspective machine, used to project images onto a grid. Source: Reprinted from Albrecht Durer, *The painter's manual* (Nuremberg, Germany: Anton Koberger, 1525).

Many Worlds

The machine aesthetic has not been uniformly successful even when applied to the pristine world of physics. Perhaps its greatest failing has been with quantum mechanics—which, as discussed in Chapter 4, seems to be missing a few nuts and bolts. According to the Copenhagen interpretation, objects such as an electron are characterized by a wave function which represents a multiplicity of states. For example, the electron might stay in its orbit around an atom or it might jump to a higher level. When the system is observed, the wave function collapses to a single state—the electron jumps or it doesn't. Such effects are usually noticeable only at the quantum level, but in principle they can be made to apply to any object, such as Schrödinger's cat (discussed earlier).

As the American physicist Hugh Everett pointed out in 1957, this theory doesn't really stand up to analysis.[19] Why is the evolution of the wave function deterministic but the outcome of a measurement random? What mechanism causes the wave function to collapse? What counts as an observer? He argued that the problem with the theory was that it lacked the courage of its own convictions. Wave function collapse was an artificial add-on. Form and function were not as one.

Everett argued that only the wave function itself was real, and that it never collapsed. Instead, each quantum event represented a branching point in which the universe divides into two or more versions. In one copy of the universe, the electron jumps; in the other, it doesn't. Because such events are happening all the time, an implication is that there are a large and growing number of universes all playing out in parallel with one another.

Everett was motivated in large part by aesthetics—in particular, "an ideal of unity, simplicity and completeness," as one paper notes.[20] The Copenhagen interpretation was too messy, artificial, and incomplete. His "many worlds" hypothesis, in contrast, followed the equations to their logical conclusion. It took the quantum model and ran with it. The theory obviously raised questions about the nature of reality and identity, especially since it implied that each person could exist in multiple universes simultaneously, doing different things. It was perhaps for this reason that in his work Everett only referred to "reality" in quotation marks. Paradoxically, his desire for unity had resulted in multiplicity without boundaries. Unifying the equations took priority, even if it meant making infinitely many copies of the actual physical universe.

After perhaps the Copenhagen interpretation or its variants, Everett's theory is the most popular interpretation of quantum mechanics among physicists (at least in this universe).[21] According to physicist David Deutsch, "It is not some troublesome, optional interpretation emerging from arcane theoretical considerations. It is the explanation—the only one that is tenable—of a remarkable and counter-intuitive reality."[22] Physicist Frank Tipler called it "the most revolutionary, beautiful, elegant, and important idea to be advanced in the past two centuries."[23] And a similar idea has also been harnessed more recently to explain how we came to exist here at all. Our universe is not alone in the world. Instead, it is just one of a whole series of mechanistic universes, churned out by a universe factory.

The Cosmic Put-Up Job

One of the great puzzles of the universe is that, as with a living system, it appears to have been very finely tuned. Consider, for example, stars, which act as factories of any element larger than hydrogen or helium. When the stars die, they expel those elements out into space to form the building blocks of other things, including us. As they say, we are all made of stardust.

Stars typically form when a gas of protons and electrons condenses under the force of gravity. As they come closer together, the positively charged protons repel each other. The nuclear strong force is not quite strong enough to overcome this repulsion, so in order to create atoms some of the protons must first transform into neutrons. This is a slow process, controlled by the weak force, and it prevents the star from burning out too quickly. Nuclear reactions necessary for the formation of larger elements also depend on the subtle dynamics between the attractive strong force and the repulsive electromagnetic force.

The whole star-making procedure therefore represents a fine balancing act. If the gravitational force were too weak, then stars and galaxies would not have condensed properly in the early universe; but if it were too strong, they would have collapsed. If the strong force were a little different, as the British astronomer Fred Hoyle showed in 1953, it would upset the balance of nuclear reactions responsible for the formation of carbon—not good for carbon-based life forms. If the balance between gravity and the electromagnetic force were off, then stars would be on average too hot or too cold to do their job. If the weak force were different, the ratio of neutrons to protons would have changed, again with catastrophic effects on nuclear reactions and the supply of elements.

Physicists complain that they have to carefully adjust the parameters in the Standard Model, such as particle masses and field strengths, to make it work, but it looks like someone has already been there before them. Indeed, the more closely you look at the universe, the more contrived it seems to be. Water, for example, has the special property that its frozen form is lighter than its liquid form. The layer of ice on top of a lake in winter shields the life forms underneath. If the electromagnetic force were a little different, then water would freeze from the bottom up and frozen lakes could not support life.

The most spectacular piece of cosmic fine-tuning is in the energy of the quantum vacuum. According to quantum field theory, empty space is actually full of virtual particles that are constantly bubbling in and out of existence. This produces a huge amount of energy, which through Einstein's $E=mc^2$ is equivalent to about 10^{93} grams per cubic centimeter (dark energy is negligible in comparison). According to general relativity this energy should contribute a negative pressure term to Einstein's cosmological constant and thus make the universe expand. However, observations show that the net effect is actually extremely small. The calculations from quantum field theory are off by a factor of about 10^{120}, which Stephen Hawking called the biggest failure of a scientific theory in history.[24] It appears that the various components of vacuum energy have been cunningly arranged so that they cancel out. This in turn allows the universe to expand quickly enough that it doesn't collapse back in on itself, but slowly enough that galaxies have time to form.

Finally, there is the question of how the cosmos blew up during the big bang in order to give a highly homogeneous distribution of matter throughout the visible universe. The explosion seems to have been controlled to an incredibly high degree. The usual explanation, first proposed in 1980 by Alan Guth and independently by Katsuhiko Sato in 1981, is as follows: during the first 10^{-34} seconds or so following the big bang, the universe underwent a period of extremely rapid expansion known as cosmic inflation (the choice of name may have had something to do with the fact that the inflation rate in the United States reached nearly 15 percent in 1980).[25] This stretched out the universe in such a way that it smoothed out the distribution of matter. The part that is visible to us could be an insignificant fraction—one part in 10^{10} or less—of the total universe. However, while cosmic inflation provides what Roger Penrose (a critic of the theory) calls "an aesthetically pleasing physical picture," the theory requires new Higgs-like particles to produce the inflation, and it also requires fine-tuning to create the right sort of result.[26]

As Fred Hoyle noted, it all looks like a "put-up job."[27] We are used to this kind of fine-tuning for living things. We have all marveled at how a human embryo turns step by step into a baby, which then grows to adulthood—everything just falls into place, with the right number of arms and legs and so on in their proper locations, as if carefully designed and controlled. According to the theory of evolution, this apparent miracle is just the product of many generations of natural selection—embryos that don't develop properly don't go on to produce progeny, so their faulty genes get filtered out. But how can we explain this for the universe?

One approach, which has been around in different forms for about three decades and is gaining in popularity, is the multiverse theory. According to this theory, there are a multitude of universes being created. Most of them don't survive or go on to produce life. And in place of natural selection, there is another selection mechanism—us.

The Ultimate Ensemble

The multiverse theory comes in various forms and flavors, but nearly all agree that the period of cosmic inflation that our universe went through after the big bang was not a unique event. Indeed, it is happening all the time—and new universes are constantly budding off, at vast distances from one another, from what physicist Andrei Linde called an eternal chaotic inflating field.[28] According to perhaps the most general or extreme version, the Ultimate Ensemble theory of the MIT cosmologist Max Tegmark, each universe is not just characterized, but is actually *defined* by a unique set of mathematical equations; conversely, any set of mathematical equations can have a corresponding universe. Our own universe is defined by a certain set of unified physical laws—a perfect model—which once discovered will fit on a T-shirt, as these things do. In most other universes, the cosmic weather is too harsh to support life, but on those that do, "the T-shirts on sale mostly have different equations."[29]

Paul Dirac said that "God is a mathematician of a very high order, and He used very advanced mathematics in constructing the universe."[30] But actually, it looks like the creator is an ensemble forecaster. In ensemble forecasting, perturbations are made to the initial conditions and the model parameters in the hope that the distribution of forecasts will in some sense "include the truth." According to the Ultimate Ensemble theory, universes are constructed in the same way—by randomly varying the model, which here is the same as reality, in the hope that it eventually works.

As discussed in Chapter 6, one feature of superstring theory is its flexibility, which means that it can be used to explain just about anything. Strings live in multiple dimensions, some of them compactified so we don't notice them. The low-energy physics of our world—including everything from the speed of light to the mass of an electron—is determined by the geometry of the compactified extra dimensions. While this flexibility is usually considered a drawback, in multiverse theory it becomes an asset, because it leads to more choices. In other universes, the extra dimensions might have curled up in different ways—or not at all. Quantum fluctuations at the birth of each universe would lead to even more variability. The Ultimate Ensemble Forecaster would have no shortage of material to work with.

So how is it that we happened to end up in a habitable universe when most universes would presumably either fail completely or be doomed to sterility? The answer, according to the anthropic principle proposed by astrophysicist Brandon Carter in the 1970s, is simply that in order to be observed, a universe must support life. All the other universes that don't support life are never observed. Furthermore, as Linde points out, the ensemble theory explains why it is that scientists can make accurate predictions based on elegant equations: "Rapid development of the human race was possible only because we live in the part of the multiverse where the long-term predictions are so useful and efficient that they allow us to survive in the hostile environment and win in the competition with other species. . . . We can only live in those universes where the laws of physics allow our existence, which requires making reliable predictions. In other words, mathematicians and physicists can only live in those universes which are comprehensible and where the laws of mathematics are efficient."[31] The best evidence for the beautiful mathematical nature of the cosmos is . . . mathematicians. They can prove it just by existing.

In *The God Delusion*, Dawkins describes the multiverse theory as beautiful because of the way it turns a functioning universe into a logical necessity.[32] He notes that people may find it to be somewhat profligate, but in fact "the multiverse, for all that it is extravagant, is simple . . . if each one of those universes is simple in its fundamental laws, we are still not postulating anything highly improbable."[33] According to Dawkins, "the beauty of the anthropic principle is that it tells us, against all intuition, that a chemical model need only predict that life will arise on one planet in a billion billion to give us a good and satisfying explanation for the presence of life here."[34]

One "chemical model," multiple universes. Again, this definition of "what we mean by beauty" is derived from a traditional aesthetic that favors elegance and rationality in the model over parsimony in the embodied reality; and the world of abstract Forms over the actual world. In the same way that unified theories lead to the prediction of new and undetectable particles, the quest for mathematical logic and consistency leads to the prediction of new and undetectable universes.[35] However, the living universe is not a cheap conjuring trick for which a model can give a simple and comforting explanation. The fact that I find the brute-force anthropic/multiverse/Ultimate Ensemble approach downright boring, rather than beautiful, just goes to show how subjective aesthetics can be.

The Virtual Universe

The Oxford philosopher Nick Bostrom goes further and notes that if there are uncountably many universes, then at least some of those universes would have spawned civilizations that are capable of making a simulated universe, as in the *Matrix* films. Furthermore, because simulations, once made, are easy to reproduce, it is likely that those civilizations would have made multiple copies. Some of those simulated universes might even have spawned simulated civilizations which in turn could make simulations, like a film within a film. So, logically, there should be many more simulated universes than ordinary ones, which after all require a lot of mass and energy to produce. "There is a significant probability that you are living in [a] computer simulation," concludes Bostrom. "I mean this literally . . . you exist in a virtual reality simulated in a computer built by some advanced civilization. Your brain, too, is merely a part of that simulation."[36] Nature is not just "meaningless" (Whitehead) or "pointless" (Weinberg) or "chemical scum" (Hawking)—it isn't even "real" (Everett). What you take to be your life could actually be an illegal download of some virtual teenager's computer game.

While Bostrom's approach may seem somewhat extreme, the idea that we live in a virtual universe is consistent with the drive towards viewing the cosmos as the product of abstract equations, like a piece of computer code.[37] We can decode the universe the same way we decode DNA. (What would happen if this code was discovered? Would it spontaneously generate a new universe in a computerized big bang? Could we download it onto our mobile phone and listen to it? Could we hack into it and tinker with the tuning? Or do the equations need to be supplemented with a starter pack of matter and energy—some kind of *stuff* with which to play?)

Even thinking about the possibility that the universe is virtual is enough to bring on a Cartesian "I think therefore I am" moment, in which you begin to question whether anything is real or whether it is all just an illusion created by a machine. In his *Meditations on First Philosophy*, Descartes wrote, "if I look out of the window and see men crossing the square, as I just happen to have done, I normally say that I see the men themselves. . . . Yet do I see any more than hats and coats which could conceal automatons?"[38] Apparently he liked on occasion to do his thinking while lying in a large bread oven, perhaps because there were no windows there. As a young man, Alan Greenspan also dabbled in logical positivism. The psychologist Nathaniel Branden described their meetings as part of the group around Ayn Rand: "In his 20s he's sitting there in my apartment and saying that he cannot say with certainty that he exists."[39] This seems quite fitting since, as Chairman of the Federal Reserve, Greenspan oversaw the increased virtualization of the U.S. economy where it turned out that much of the money didn't exist either.[40]

Now, it seems to me that there are two ways to look at all this. One is that we are at a great moment in the evolution of our species and our universe. By combining insights from mathematics, particle physics, and cosmology, the human mind is on the verge of peeling back the cosmic veil to reveal the What Is (or, for multiple universes, the What Are). As a 2011 editorial in *New Scientist* put it, "Almost everything in modern physics, from standard cosmology and quantum mechanics to string theory, points to the existence of multiple universes—maybe 10^{500} of them, maybe an infinite number. . . . If our universe is just one of many, that solves the 'fine-tuning' problem at a stroke: we find ourselves in a universe whose laws are compatible with life because it couldn't be any other way."[41] Our cosmos is the inevitable product of a beautiful machine. The Pythagorean dream, to know the mathematical scheme which underlies reality, is about to be realized.

Another interpretation—which I prefer for its simplicity—is that large parts of the scientific community are suffering from a severe case of the perfect model syndrome. There is no hard evidence for string theory, or for multiverses, or even for cosmic inflation—but somehow, they have joined together in a mutually reflecting hall of mirrors to provide the standard scientific explanation for fine-tuning. The models have become confused with reality. We cannot infer anything about the existence or properties of other universes by analyzing string theory, because *that is just something we made up*.[42] It hasn't been shown to work for our universe, let alone others. Things like the multiverse and higher dimensions are fun to talk about, but they work just as well without the modern scientific trappings—as far back as in 1277 the bishop of Paris ruled it heretical to say that God "could not make several universes" or "make more than three dimensions exist simultaneously."[43] (The idea that we are living in a movie or a game, however, would certainly explain a few things. I have always been intrigued by the fact that the Moon appears to us to be exactly the same size as the Sun, which allows for perfect eclipses—and which, as discussed in Chapter 4, helped to confirm Einstein's theory of relativity. What does that mean? Is it a sign, placed there by a director? And if we do figure it all out, will the universe "instantly disappear and be replaced by something even more bizarre and inexplicable," as Douglas Adams put it?[44])

In her book *Physics on the Fringe*, about "outsider science," Margaret Wertheim found some of the crankiest at a luminary-studded conference on string theory and cosmology. Despite the admitted lack of any actual evidence for their theories, "everyone had the Answer and could barely contain their excitement about sharing it with their peers. . . . After two days, I couldn't decide if the atmosphere was more like a sugar-fueled children's birthday party or the Mad Hatter's Tea Party—in either case, everyone was high."[45] The relevant type of inflation here appears to be what Jungian psychologists called psychic inflation, a kind of ego singularity event in which a person "thinks with mounting excitement that he has grasped the great cosmic riddles; he therefore loses all touch with human reality."[46]

The multiverse theory is not universally popular amongst scientists. Some believe that the unification program will eventually show that our universe is the one possible solution to the laws of physics; such scientists tend to see the multiverse theory as a threat to their funding, which along with aesthetics is a prime criterion for theory selection.[47] But short of that, the multiverse theory offers an apparently logical and consistent mechanism for how our universe came to be produced while preserving the primacy of the mathematical model. The debate over the details of the theory preserves the impression that the underlying science is being vigorously contested, while leaving its core assumptions untouched.

Anthropologists say that the best way to understand a society is to listen to its stories and in particular its creation myths. In ours, we were made by a machine. And that means we too are machines. We are rational economic man at home in the rational economic universe. It is ironic that atomism, which aimed to dispel the need for myths and stories, became the source for the greatest myth of all.

Box 9.2

Atomic Evolution

Darwin's natural selection is a classic example of a beautiful, elegant, and powerful theory. Today, many scientists see life in Neo-Darwinist terms as a competition between genes, which are subject to random mutations. As Francis Crick put it, "Chance is the only source of true novelty."[48] But are random change and competition really the only contributors to evolution?

The biologist Lynn Margulis argued that most evolution historically took place in microorganisms and has involved a process called endosymbiosis, in which species exchange components or come together to form new species.[49] For example, cells such as those in our bodies contain mitochondria which used to be a separate organism. We also have a symbiotic relationship with the microbes in our gut, which help us to digest food. According to Margulis, "Mergers result in the emergence of new and more complex beings. I doubt new species form just from random mutation."[50]

Margulis was also associated with the Gaia hypothesis, first proposed by the English scientist James Lovelock, which is based on the observation that the planet Earth has many properties of a living organism—including regulation of the climate so that temperature remains within a band suitable for life—despite the fact that it is not subject to Darwinian evolutionary pressures.[51] Paleontologist Scott Sampson described the theory as "the deepest, most beautiful scientific explanation" of all, even if it lacks mathematical precision.[52] (By contrast, Richard Dawkins calls it "dangerous and distressing to scientists who value the truth," since there is no Darwinian reproduction or selection at work.[53])

The biologist and complex systems researcher Stuart Kauffman also points out that biological systems have a remarkable capacity for self-organization in which highly organized patterns can emerge without any planning or selection.[54] Complexity scientist Steven Wolfram similarly sees the patterns of nature emerging from internal dynamics, rather than just natural selection.[55]

Neo-Darwinism's rather post-apocalyptic-sounding emphasis on mutations and survival of the fittest is consistent with the idea, going back to Democritus, that the world is determined by the random shuffling of atoms: "Everything existing in the universe is the fruit of chance and necessity."[56] But while random mutations and selection are clearly important drivers of evolution, that doesn't grant them exclusivity. The difference between the mainstream reductionist approach and the complexity approach is revealing—not just about the natural world but also about our own values and aesthetic choices.

The multiverse theory is even more extreme than Neo-Darwinism in that it dispenses with Darwinism. Like fruit flies being bombarded with X-rays, or humans being exposed to nuclear radiation, new universes are constantly undergoing mutations to their mathematical DNA. Most of them die, but every now and then it seems that a remarkable new creature is produced, without any need for competition or selection, except by the anthropic principle. God doesn't just play dice with the universe; he plays dice with universes.

In his book *The Life of the Cosmos*, Lee Smolin argued that those universes that survive could give birth to new universes with slightly mutated versions of their own characteristics. This would restore an element of selection in the sense that workable versions would come to predominate.[57] But it is hard to see how universes could duke it out in competitive contests, or perform elaborate mating rituals, to achieve prominence—which I think is what Darwin had in mind. Nor does the theory explain how, why, or by whom the ensemble of universes is being made in the first place. Unless, of course, they are actually films in a film festival, and selection is performed by a panel of judges (the Multiplex theory).[58]

Model Grooming

In this book, I have argued that the scientific aesthetic is characterized by a number of properties—harmony, stability, unity, symmetry—which have been favored since the time of the ancient Greeks. The Ultimate Ensemble, like Neo-Darwinism (see Box 9.2), superstring theory, or for that matter our predictive models of the world, is an attempt to force nature to conform with this aesthetic. The idea that the universe is identified by a set of mathematical equations is the same as the Pythagorean notion that number is all. The idea that all change must be random—driven either by mutations or quantum fluctuations—is consistent with the calming atomistic philosophy of Democritus, which insisted that both the atoms and the rules which govern them remain the same. We live in an essentially stable, machine-like universe that is governed by, and is also in a sense equivalent to, a perfect model.

One reason why scientists cling to their models in any branch of science is simply that it takes so much time and energy to develop them. The sunk costs, to use an economics term, are huge. Models of the Earth's climate, the human body, the economy, or the universe have been painstakingly built up over decades by generations of scientists. The string theorist Michio Kaku wrote, "If string theory itself is wrong, then millions of hours, thousands of papers, hundreds of conferences, and scores of books (mine included) will have been in vain. What we hoped was a 'theory of everything' would turn out to be a theory of nothing."[59] Lee Smolin added, "It is not an exaggeration to say that hundreds of careers and hundreds of millions of dollars have been spent in the last thirty years in the search for signs of grand unification, supersymmetry, and higher dimensions. Despite these efforts, no evidence for any of these hypotheses has turned up."[60] The costs increase by at least a factor of 100 if you count the work on accelerators. Beauty, it seems, doesn't come cheap.

Even if those extra dimensions never show up, string theory has helped to stimulate new branches of mathematics. The technology behind projects such as particle accelerators has also been hugely beneficial and has helped foster international cooperation, as discussed above. But if science is being led astray by aesthetics, as I have argued here, then much scientific research carried out since the late 1960s—when the Standard Model was mostly complete—could be described as a form of grooming for the perfect model. String theory, supersymmetry, virtual humans, neoclassical economics, Ultimate Ensembles, the virtual universe—all are manifestations of the same phenomenon. To borrow an analogy from Richard Dawkins, our scientific genes have taken over our collective mind, but their only aim is to preserve themselves and keep our belief in the models alive.

Decadent Science

In his 1897 essay *What is Art?* Tolstoy considers the state of decadent fin de siècle European art. "We have the terrible probability to consider," he writes, "that while fearful sacrifices of the labor and lives of men . . . are being made to art, that same art may be not only useless but even harmful."[61] His complaint with this voluminous, expensive, yet low-quality artistic output was that it was all based on hackneyed ideas of beauty and was therefore counterfeit rather than genuine art.

To develop his argument, he first tries out several theories of aesthetics, but decides that whether beauty is seen as "an imitation of nature, or as suitability to its subject, or as a correspondence of parts, or as symmetry, or as harmony, or as unity in variety, etc.," these theories all amount to the circular definition that "art is that which makes beauty manifest, and beauty is that which pleases."[62] But this association of art with pleasure is inadequate, and Tolstoy argues that art is instead about communication: what makes a work of art effective is its infectiousness, which in turn depends on its individuality, clearness, and sincerity. Equally important is the subject matter, and this has to be aligned with the religious impulse, which Tolstoy saw as the central driving force of a society. Questions about art therefore come down to questions of the communication of religious values. Beauty is a secondary issue; a distraction.

Tolstoy's definition of religion was quite broad: "In every period of history, and in every human society, there exists an understanding of the meaning of life which represents the highest level to which men of that society have attained, an understanding defining the highest good at which that society aims. And this understanding is the religious perception of the given time and society. . . . Such a religious perception and its corresponding expression exists always in every society. If it appears to us that in our society there is no religious perception, this is not because there really is none, but only because we do not want to see it."[63] He described the religious perception of his time, in its "widest and most practical application," as the consciousness that "our well-being, both material and spiritual, individual and collective, temporal and eternal, lies in the growth of brotherhood among all men—in their loving harmony with one another."[64]

While Tolstoy's theory is controversial—he concluded for example that Beethoven's ninth symphony was bad art—his arguments could apply quite well to science, which in many quarters of Western society plays a role similar to that of religion, at least according to Tolstoy's definition. Its cathedrals are universities and scientific facilities like particle accelerators; its priests are scientists; its mystics are the long-haired and wild-eyed visionaries who enliven science documentaries; its bible is the laws of physics. To quote Einstein: "A contemporary has said, not unjustly, that in this materialistic age of ours the serious scientific workers are the only truly religious people."[65]

To many people, it is science that seems to hold the answers to the big cosmic questions. Who are we? Where do we come from? Where are we going? And it is science that provides a sense of beauty and wonder. The strength of this identification lies in the power and mystique of scientific models, and the equation of what biologist Antoine Danchin calls the truth of the model and the truth of the phenomenon: "It is a common confusion between these two kinds of truth—the norm in magic—that sometimes sanctifies the model (which is regarded as part of the real world) and gives the scientist the role of priest."[66] (A difference with religions, of course, is that scientists are supposed to put their ideas to the test, though this distinction fades somewhat when theories are not falsifiable.)

By Tolstoy's criteria, the question of whether a scientific work is considered good depends not just on what it tells us about the world, but also on how it communicates scientific values. The traditional scientific sense of aesthetics is related to ideas about harmony, symmetry, unity, and so on—but, as Tolstoy noted, these are strongly held but ultimately sterile attempts to define beauty. What are more interesting are the *values* which find expression in scientific works and which aesthetic choices reflect. These are essentially the Apollonian virtues of clarity, rationality, detachment, power, order, and prediction. (Note that, unlike with most religions, these ideas about what is good have little to say about ethics, outside of professional standards such as honesty: the nuclear bomb was "good science" even though its consequences were horrifying.[67]) Theories that emphasize things like interconnectedness, subjectivity, and real uncertainty—of the sort that cannot be smoothed away by statistics—do not score highly.

The difference in aesthetics between the reductionist and complexity approaches is therefore about more than ideas of beauty; it is a manifestation of profoundly different but complementary visions of science. While the tensions between them are being worked out, significant portions of the scientific work currently being produced, often on an industrial scale, are just paying homage to old aesthetic standards. The mainstream approach to questions such as the economy, the climate, and even the birth of the universe is similar to Tolstoy's counterfeit art: it looks like the real thing but it smells like a fake.[68]

This shouldn't put people off science—quite the opposite. As discussed in the final chapter, the future of science has never looked brighter. We just need to change hands for a while.

10

The Left-Hander's Guide to the Cosmos

But science, spurred by its powerful illusion,
speeds irresistibly towards its limits where its optimism,
concealed in the essence of logic, suffers shipwreck. For the
periphery of the circle of science has an infinite number of
points; and while there is no telling how this circle could ever
be surveyed completely, noble and gifted men nevertheless
reach, e'er half their time and inevitably, such boundary points
on the periphery from which one gazes into what defies
illumination. When they see to their horror how logic coils
up at these boundaries and finally bites its own tail—
suddenly the new form of insight breaks through,
tragic insight which, merely to be endured,
needs art as a protection and remedy.

Nietzsche, *The Birth of Tragedy*

The ostensible scientific view is that life is a
chemical reaction. . . . The corresponding mystical view is
that life is a beautifully unknowable thing. . . . Between these
extremes we have the profoundly important, but poorly
understood, idea that the unknowability of living things
may actually be a physical phenomenon.

Robert B. Laughlin, *A Different Universe*

The universe is made of stories, not of atoms.

Muriel Rukeyser, *The Speed of Darkness*

The left brain (which controls the right side of the body) favors rational intellect over intuition, abstract thought over concrete details, linearity over complexity, stability over change, analysis over a holistic approach, and objectivity over subjectivity. In aesthetic terms, it favors simple symmetry and straight lines. To say—in a rational, objective kind of way—that science tilts in favor of the left brain would not be a risky statement. What would the implications be for different areas of science, including physics, biology, and economics, if our scientific quest for beauty engaged both sides of our brain? This final chapter argues that by adopting a new scientific aesthetic, we may come to a greater understanding of our place in the universe.

Quantum physics is full of paradoxes. For example, there is the double-slit experiment where photons pass one at a time through the slits as if they were individual particles, but then collectively over time build up an interference pattern as if they were waves. If one slit is closed, the interference pattern disappears. The photons seem to have the ability to scan the terrain and count the slots before they move. Or there is the paradox of Schrödinger's cat, described in Chapter 4, where a cat can be both alive and dead at the same time.

Niels Bohr once said, "How wonderful that we have met with a paradox. Now we have some hope of making progress."[1] The solution of the Copenhagen model was to arrive at a rigorous, consistent, mathematical formalism which was capable of matching all observed experimental data to a high level of precision. But some people continued to be troubled—especially Einstein, who for years tried to invent thought experiments which would prove that quantum physics was an incomplete theory.

In 1935 Einstein co-wrote a paper that presented what became known as the Einstein-Podolsky-Rosen paradox (EPR).[2] This arises when two quantum systems are related in such a way that information on one yields information on the other, in a manner that appears to contradict the uncertainty principle. For example, if an electron and a positron are emitted during the decay of a pion, then the electron's spin can be measured around one axis and the positron's spin can be measured around another axis.[3] Because the particles were created from the same source, the spin on one tells us the spin on the other. That appears to contradict the uncertainty principle, which says we can only get accurate measurements of the spin around one axis at a time.

The only way out of this paradox is to assume that the particles could somehow communicate, so that a measurement on one instantaneously affects the other—an effect known as quantum entanglement, or to use Einstein's phrase, "spooky action at a distance."[4] The EPR paper therefore concluded that quantum theory was not a complete theory of reality, but only the statistical approximation of some deeper theory based on "hidden variables" that would restore causality.

In 1964, however, the Irish physicist John Bell devised an ingenious way to distinguish between quantum mechanics, with its non-local properties, and Einstein's hidden variables.[5] He showed that if one took repeated measurements for a large number of events, then the statistics predicted by the two models would be different. Since then, numerous laboratory experiments analogous to EPR have been carried out and quantum mechanics has been repeatedly vindicated. The universe, it seems, is spooky—and entangled.

The Quantum Mind

It is something of a cliché to say that quantum physics is so weird that it is beyond our understanding (one almost suspects that scientists *want* it to be seen as baffling).[6] Things like duality, uncertainty, and entanglement cannot be handled using everyday logic. But is the problem with quantum physics—or is it with logic?

Philosophers have been struggling with paradox since philosophy was invented. As discussed in Chapter 2, Zeno argued that movement was impossible on the basis that in order to pass from one point to another, an object would have to travel half of the way, and then half of the rest of the way, and so on indefinitely, thus it could never arrive at its destination.

Indeed, if one were to write a computer program that broke movement down into a succession of individual steps, like the still pictures that make up a film, but with the steps getting progressively smaller as

Zeno describes, then the program would enter an "infinite loop" and run forever (or until the programmer got bored and interrupted it). So Zeno's paradox is only a paradox if we think in the same way as a computer, that is, in terms of a sequence of static events rather than dynamic flow.

Another famous paradox was due to Epimenides, a Cretan who declared that "all Cretans are liars." Again, if you interpret this statement in a computer-like fashion, then it leads to a paradox because Epimenides is a Cretan, so his statement is false, so all Cretans are not liars, so the statement is true, and so on. Most humans, in contrast, would take the context into account and realize that Epimenides is making an exaggerated statement about *other* Cretans, but maybe flag up a mental warning that he is a Cretan himself.

In the same way, the paradoxes of quantum physics do seem illogical according to the law of cause and effect. But it doesn't follow from this that the subatomic world is immune to understanding—at least in a looser, fuzzier kind of way than that favored by quantum mechanics. Indeed, there is plenty of evidence to show that the human mind works in a rather similar manner. As physicist Diederik Aerts notes, "People often follow a different way of thinking than the one dictated by classical logic. The mathematics of quantum theory turns out to describe this quite well."[7]

For example, in the same way that photons behave differently when presented with two slits or just one, our decisions can be swayed in strange ways by the available options. In one experiment, psychologists offered subjects a game in which they had an even chance of either winning $200 or losing $100.[8] After playing once, they were offered the chance to play again. If they were told that they won the first game, 69 percent decided to gamble their winnings. If they knew they lost, then 59 percent tried to make back their losses. But if they were not told the result, then only 36 percent opted to repeat. Why the difference? It is as if the simultaneous possibility of the two outcomes—winning and losing on the first game—creates an interference pattern in the mind which affects the decision-making process.

Quantum physics is context-sensitive, in that the answer we get by interrogating a system will depend on the question that is posed. If we measure momentum, then position becomes fuzzy. We can even modify the behavior of a particle just by checking to see if it has moved. Context is also important in human reasoning, which is why our brains don't go into infinite loops when interpreting statements like the Cretan paradox. The strongest demonstration of the power of framing is the placebo effect, where a sugar pill can become a medicine just by being presented as one. Background context curves the lines of linear logic in the same way that gravity curves lines in spacetime.

Behavioral economists have shown that when making economic decisions, logic—of the sort that characterizes cause-and-effect classical physics—actually plays a rather small role.[9] We make choices based on a range of cognitive processes, which then "collapse" to a single decision in a manner similar to wave function collapse—though if queried, we then tend to justify the decision based on a logical argument. The insights from behavioral economics have been controversial amongst mainstream economists but have proved popular with people like marketers, who are happy to exploit the fact that we are easily swayed by context, framing, and emotions.

One aspect of quantum physics, which its founders struggled with, was the idea of discontinuous change. Particles could only exist in certain allowed states, so for example an electron might suddenly jump from one orbit around a nucleus to another without occupying intermediate states along the way. This seems to contradict our basic ideas about movement; however, at a psychological level, we are all familiar with the "Aha!" moment when a solution to some puzzle seems to spring into our mind fully formed. Many scientists including Kepler, Newton, Mendeleev, and Einstein reported that their greatest insights came in the form of a sudden epiphany, rather than as the linear outcome of incremental changes. Creativity, it seems, relies on quantum leaps.

Viewed in this way, the ideas of duality, discontinuity, uncertainty, and entanglement no longer seem quite so bizarre. Indeed, perhaps the weirdest thing about the theory of quantum physics is that it has presented these properties to us as being incomprehensible features of some alien landscape, while in fact they have been there all along—in our own brains.

The Dual Brain

One of the more obvious and incontrovertible aspects of the human brain is its asymmetry. The brain is divided into two hemispheres, which have physically distinct properties and are separated by a large bundle of nerve fibers known as the corpus callosum. The left hemisphere controls the right-hand side of the body and vice versa. In the healthy brain, the two hemispheres are in constant communication, so there is no clear division in their function. However, it has long been realized that they play different roles in cognition.

The English word "left" is from the Anglo-Saxon word *lyft* for weak, perhaps because most humans are right-handed. This association of the left side with weakness has carried over to the right side of the brain. When scientists began to investigate brain lateralization in the mid-nineteenth century, the right brain was considered by many to be not only weak, but practically dead. In 1926, Salomon Henschen wrote, "In every case the right hemisphere shows a manifest inferiority when compared with the left, and plays an automatic role only. . . . The question therefore arises if the right hemisphere is a regressing organ . . . it is possible that the right hemisphere is a reserve organ."[10] It was only in the 1960s, with the work of the psychologist Roger Sperry and his collaborators, that the right brain was found to be anything more than a kind of mute companion to the left.

Through experiments on "split-brain" patients, in which the connection between the two hemispheres had been deliberately severed to prevent epileptic seizures, Sperry found that the two hemispheres of his patients were carrying on quite separate existences. The left brain was "highly verbal and mathematical, performing with analytic, symbolic, computerlike, sequential logic."[11] However, "the so-called subordinate or minor hemisphere, which we had formerly supposed to be illiterate and mentally retarded and thought by some authorities to not even be conscious, was found to be in fact the superior cerebral member when it came to performing certain kinds of mental tasks. The right hemisphere specialties were all, of course, nonverbal, nonmathematical and nonsequential in nature. They were largely spatial and imagistic, of the kind where a single picture or mental image is worth a thousand words." These right-hemisphere tasks included face recognition and pattern detection.

The idea that left brains and right brains are different was of course immediately hijacked—by the left brain—with the result that a highly simplified version became part of pop culture. Accountants were left-brainers, artists were right-brainers. A scientific backlash then ensued in which psychologists and neuroscientists pointed out that the two hemispheres share many tasks and work together, so it is illogical—in a left-brained way—to say that any person is left-brained or right-brained.

In the meantime, though, Sperry's work has been complemented by a long series of experiments that have helped to further tease apart the contrasting roles played by the two sides. These experiments involve patients with damage to one hemisphere but not the other; patients with damage to the corpus callosum; or healthy subjects, with images projected to the separate hemisphere of each subject via the corresponding eye. As a whole, this body of work has built up a picture of the separate hemispheres as being almost two separate personalities. In the same way that matter has a particle aspect and a wave aspect, the two hemispheres of the brain have complementary natures.

It is obviously dangerous to overgeneralize; everyone's minds and personalities are different, and so are everyone's left and right brains. Functions that are handled by one hemisphere for most people may cross over to the other hemisphere for others.[12] What we know about the hemispheres comes from experiments that attempt to artificially pry apart their functions, so we know more about their behavior under stress than in healthy conditions. As psychologist Marie Banich notes, "Interhemispheric interaction has important emergent functions—functions that cannot be derived from the simple sum of its parts."[13]

With these caveats in mind, if we want to understand how our minds work, it still seems a no-brainer to explore the differences between the left and right brains. Continuing with this book's theme of lists of opposites, another such list appears in Iain McGilchrist's book *The Master and his Emissary*, where he characterizes the different contributions of the two hemispheres in terms of a number of opposing principles. These should be considered as tendencies rather than as absolute statements. The list includes:

"Whole versus part." The right hemisphere takes a holistic view while the left hemisphere sees systems in terms of their component parts. Drawings by people with damage to the right hemisphere break objects down into a collection of unrelated parts with no sense of the whole.

"Context versus abstraction." The right hemisphere takes context and relationships into account. The left hemisphere prefers abstract symbols such as numbers, is more narrowly focused, and is insensitive to context.[14]

"Reason versus rationality." The right hemisphere integrates different types of knowledge and emotions into a reasoned (as opposed to purely rational) judgment, and it is open to paradox. It also plays the role of "bullshit detector."[15] The left hemisphere favors narrow logic and is adept at constructing complicated models, but is vulnerable to "naively optimistic forecasting of outcomes."[16]

"Possibility versus predictability." The right hemisphere deals with what is new and unexpected. Like a quantum wave function, it can "hold several ambiguous possibilities in suspension together without premature closure on one outcome."[17] The left hemisphere seeks stable patterns in transitory phenomena in an effort to make them predictable and controllable. Its drive is "towards manipulation, and its ruling value is utility."[18] It will tend to confabulate a rational justification if proven wrong. It "needs certainty and needs to be right."[19]

"Living versus non-living." The right hemisphere engages with anything that is considered to be alive, while the left hemisphere deals with machines that can be understood in terms of their parts. The right hemisphere "will empathise with, identify with, and aim to imitate only what it knows to be another living being, rather than a mechanism" (an exception is musical instruments).[20] In contrast, "The model of the machine is the only one that the left hemisphere likes. . . . The machine is something that has been put together by the left hemisphere from the bits, so it is understandable purely in terms of its parts; the machine is lifeless and its parts are inert."[21]

As McGilchrist notes, the two hemispheres see, experience, and respond to the environment in very different ways: "The world of the left hemisphere, dependent on denotative language and abstraction, yields clarity and power to manipulate things that are known, fixed, static, isolated, decontextualized, explicit, disembodied, general in nature, but ultimately lifeless. The right hemisphere, by contrast, yields a world of individual, changing, evolving, interconnected, implicit, incarnate, living beings within the context of the lived world."[22] If somehow the left hemisphere became dominant, then the world "would be relatively mechanical, an assemblage of more or less disconnected 'parts'; it would be relatively abstract and disembodied; relatively distanced from fellow-feeling; given to explicitness; utilitarian in ethic; over-confident of its own take on reality, and lacking insight into its problems—the neuropsychological evidence is that these are all aspects of the left hemisphere world as compared with the right."[23]

The left hemisphere tends towards optimism and believes in its own theories. It is highly competitive and "its concern, its prime motivation, is *power*."[24] When things don't work out as expected, it shows: "Denial, a tendency to conformism, a willingness to disregard the evidence, a habit of ducking responsibility, a blindness to mere experience in the face of the overwhelming evidence of theory."[25] It creates a "self-reflexive virtual world" like a "hall of mirrors" from which it cannot escape unaided.[26]

The two hemispheres also have different aesthetic sensibilities. The left brain is drawn to straight lines and to symmetry, which represents a kind of visual simplification.[27] A 1988 study by Regard and Landis, which projected images to the separate hemispheres of their subjects, found that the left hemisphere likes its pictures to conform to the gestalt law of Prägnanz—which tends towards meaning, simplicity, completeness, and symmetry—while the right hemisphere prefers unusual and incomplete forms.[28] In her classic art instruction book *Drawing on the Right Side of the Brain*, Betty Edwards teaches students of drawing how to access the right brain in order to avoid the abstract, lifeless forms produced by an overreliance on the left brain.[29]

The Tragedy of Physics

As I have argued elsewhere, science has traditionally been based on a left-brained, theoretical approach that considers holistic or systems thinking to be dangerously woolly.[30] The philosopher Mary Midgley wrote that "during much of the twentieth century the very word 'holistic' has served in some scientific circles simply as a term of abuse."[31] This trend is particularly strong in economics, which after all is the area that invented the concept of rational economic man. The neoclassical economist Frank Knight wrote in 1921 that "this is the way our minds work; we must divide to conquer. Where a complex situation can be dealt with as a whole—if that ever happens—there is no occasion for 'thought.'"[32] The Pythagorean list of opposites, with its preference for the straight, the square, the right side, reads like a terse declaration of intent from the left brain.

The neuroscientist Tim Crow wrote that "except in the light of lateralisation, nothing in human psychology/psychiatry makes any sense."[33] The same applies to science, which is after all a human pursuit. In her book *My Stroke of Insight: A Brain Scientist's Personal Journey*, Jill Bolte Taylor described the sensation she felt after a stroke crippled the left hemisphere of her brain. As a scientist she had always relied on her left brain for its phenomenal ability to "categorize, organize, describe, judge, and critically analyze absolutely everything."[34] But after the stroke, she felt an enormous sense of connection with the rest of the universe: "My left hemisphere had been trained to perceive myself as a solid, separate from others. Now, released from that restrictive circuitry, my right hemisphere relished in its attachment to the eternal flow. I was no longer isolated and alone. My soul was as big as the universe."[35]

When we say that quantum concepts like duality, uncertainty, and entanglement are confounding to our brains, what we really mean is that they are confounding to our left brains. For the right brain, they are just another day at the office. The aesthetic of science can therefore be described as predominately left-brained, or right-handed.[36] Like

Renaissance painters, whose calculations of perspective were based on a single monocular viewpoint, we are looking at the world with one eye closed.

Ironically, it was probably a right brain that came up with the whole quantum idea in the first place. Einstein described his creative process thus: "The words or the language, as they are written or spoken, do not seem to play any role in my mechanism of thought. The physical entities which seem to serve as elements in thought are certain signs and more or less clear images which can be 'voluntarily' reproduced and combined. . . . The above mentioned elements are, in my case, of visual and some of muscular type. Conventional words or other signs have to be sought for laboriously only in a secondary stage, when the mentioned associative play is sufficiently established and can be reproduced at will."[37] That kind of nonverbal reasoning has been shown to be a right-brain activity.

Another specialty of the right brain, apart from sudden insight into the structure of the universe, is humor.[38] Many jokes rely on the discrepancy between mechanistic thinking and the surprise of the unexpected. They usually start by setting up (or, in the case of stereotypes, by assuming) a pattern, which we expect to continue; but then a sudden twist produces a mental spark that manifests as laughter.

Nietzsche described the tragedy of science as the moment when "logic coils up . . . and finally bites its own tail." This is when science "needs art as a protection and remedy."[39] Perhaps it also needs a sense of humor (as Carol Burnett quipped, "comedy is tragedy plus time"[40]). I sometimes suspect that the appearance of simple and elegant patterns in nature, such as the geometrical arrangements of the Standard Model, are all the setup of an elaborate cosmic joke, leading us on to a punch line which is yet to come.

The Will to Power

In some ways, it seems natural and appropriate that science should be aligned with left-brained values and aesthetics, even if its means are often right-brained. The aim of traditional science, after all, is to look for patterns in data and find elegant equations with the maximum explanatory power.[41] Areas such as particle physics seem especially suited to this approach (and may become less influential as an aesthetic guide for the rest of science as a result). The problem comes when the approach is misapplied, or taken to an extreme wherein the left-brained love of theory and simplicity becomes detached from reality. Right-brain damage is associated in individual humans with profound and unsettling effects. For example, patients may deny that the left side of their body belongs to them, or they may insist that it has been taken over by a robot (rather like Stanley Kubrick's Dr. Strangelove character, whose uncontrolled hand occasionally tries to strangle him). In the same way, many areas of science have become excessively left-brained, which hampers their effectiveness.

The history of science, up until the early twentieth century, was one of a linear buildup of knowledge in which theory and experimental data worked hand in hand in a successful way. Chemists and physicists encountered patterns in nature—the octave-like structure of the elements, the symmetries of forces—which prompted them to search out unified laws of nature. At first there was rapid progress, with scientists reaching right to the heart of the atom. The discovery of relativity and quantum theory challenged this linear, mechanistic narrative and was a clear opportunity to develop a new kind of science, modeled after Bohr's complementarity. However, insights into the quantum nature of reality were quickly hijacked and neutralized by the Copenhagen interpretation, and by the work of brilliant but hyper-rational scientists like John von Neumann and Paul Dirac. The result was a theory that was tremendously clear and successful from a narrowly computational point of view, but was otherwise unsatisfactory and alienating.

The role of scientists after the war was settled by the detonation of the first nuclear devices—and the ensuing risk that the entire planet might end up solarized. The left hemisphere's love of power found its natural ally in the military—and predominately male—pursuit of advanced technology. The relationship between brain lateralization and gender is weak and not straightforward, but the hemispheres' connection with gender roles in society is much better defined—especially as they existed in the 1950s. Scientific funding flowed into prestigious, male-dominated, abstract areas like high-energy physics. Other areas of science, such as biology (selfish gene theory) and economics (rational economic man), also celebrated left-hemisphere values such as competition and rationality. The inherent uncertainty of real-world systems was hidden away by aping the hard, probabilistic approach favored in quantum mechanics.

By the 1970s, physicists had arrived at their Standard Model, which successfully simulated subatomic experiments. But from then on progress suddenly became more difficult, as if impeded by a sticky morass of Higgs bosons. The existence of some predicted particles was confirmed—most notably the famed boson itself—but little truly novel was added (as discussed in Chapter 7, it is hoped that the LHC will yet produce some surprises). The lack of new data did not prevent theorists from extrapolating their theories wildly, however. While earlier science—Galileo rolling bronze balls down a ramp, Curie with her glowing tubes, Rutherford's students counting scintillations in a darkened room, even Einstein's "happiest thought" about gravity—had a kind of physicality and was based at least in part on bodily sensations, theory in the late twentieth century became increasingly disembodied and freed from the constraints of experiment.[42] Supersymmetric particles were evoked to restore beautiful symmetry to a rather messy model. String theory probably says more about the Pythagorean desire for numerical harmony than about the real world; and its adherents exhibit the conformism, love of abstract theory over data, and lack of functioning bullshit detector that characterizes the unchecked left hemisphere. The idea that we live in a multiple or virtual universe, meanwhile, bears some resemblance to symptoms of extreme right brain dysfunction.[43] To attain an apparently consistent and unified picture of the cosmos, we play havoc with its embodied reality, either denying it or multiplying it without bounds.[44]

In an interview, the mathematician Benoît Mandelbrot compared mathematics to Antaeus, the son of Gaia, who "had to touch the ground every so often in order to reestablish contact with his Mother; otherwise his strength waned. To strangle him, Hercules simply held him off the ground."[45] Science similarly needs some occasional "down-to-Earth input" to keep in touch with the real world. When the Trinity scientists leaped into the air to celebrate what journalist William Laurence called "the birth of a new force that for the first time gives man means to free himself from the gravitational pull of the earth," it seems they never came down. Since the late 1960s, though, a more Earthbound vision of science has been emerging that is based on a fundamentally different aesthetic and has less to do with machines than it does with living systems.

The Janus Phenomenon

In the same way that electrons or sound phonons can be seen as waves or particles, living systems are structured around a hierarchical sequence of modules, which also have a dual atomistic/holistic nature. For example, the human body contains individual cells, which have their own intricate construction. The cells are organized into structures such as organs (from the Latin word for musical instrument). We in turn organize into social units, from families to communities to nations. As the author and journalist Arthur Koestler noted in 1969, these organizational modules—which he called holons—are "self-regulating open systems which display both the autonomous properties of wholes and the dependent properties of parts."[46] Each has a "dual tendency to preserve and assert its individuality as a quasi-autonomous whole; and to function as an integrated part of an (existing or evolving) larger whole." In humans, Koestler believed, it is the integrative tendency that results in "flexible adaptations, improvisations, and creative acts which initiate new forms of behaviour." Koestler referred to this dual property as the Janus phenomenon, named after the two-faced God whose twin visages, which probably represented the Sun and the Moon, adorned many Roman coins.

Other examples of these hierarchically structured organic systems, known as complex adaptive systems, include the growth of a city, the Internet, the economy, the biosphere—and perhaps the universe.[47] A characteristic they share is that each organizational level shows emergent properties which cannot be deduced from the level below. Society is not a larger version of an "average" human any more than an ant colony is just a larger version of an ant. These systems also tend to be characterized by nonlinear feedback loops, which combine the ability for rapid response with the need for regulation; and complex networks, with multiple channels of communication between components. Their study at centers such as the Santa Fe Institute, founded in 1984 by a group of scientists including Murray Gell-Mann, has driven the development of areas of applied mathematics including complexity theory, nonlinear dynamics, and network theory.

Since the focus is on organizational principles rather than reductionism, these methods can be applied at any scale. One way to get a sense of how they differ from classical techniques is to relate them to brain lateralization. It is as if our collective right brain has suddenly fired up and decided to engage more fully in science, and not just in the role of muse, inspiration, or subservient partner.

Whole versus part. Complexity science studies emergent properties which develop from, but cannot be reduced to, simpler components. Nonlinear dynamics studies systems in which feedback loops between components can have unexpected effects at the global level. Network theory studies complex networks, such as networks of interacting genes and proteins in the human body, electrical power networks, or the financial network. In systems biology, what counts is less the "selfish genes" than how they interact. These techniques look at the connected whole rather than just the separate parts.

Context versus abstraction. Because complex systems cannot be reduced to their components, theories and models are best viewed as patches which capture only one side or aspect of the more complicated reality. One implication is that models are not easily transportable from one context to another. This is true for areas such as economics, but it also applies in physics. As physicists Nigel Goldenfeld and Leo P. Kadanoff wrote, "Up to now, physicists looked for fundamental laws true for all times and all places. But each complex system is different; apparently there are not general laws for complexity. Instead, one must reach for 'lessons' that might, with insight and understanding, be learned in one system and applied to another. Maybe physics studies will become more like human experience."[48]

Reason versus rationality. Traditional science works by breaking a system down into its component parts, figuring out the linear cause-and-effect relationships which govern it, and building a mechanistic model.

Its power comes from the clarity of its logic. With complex systems of the sort encountered in biology or economics, that approach only goes so far, and it must be combined with experience and insights based on attentive observation of the system. Theoretical idealizations such as rational economic man are not just a poor approximation to reality; they are highly misleading and potentially dangerous.

Possibility versus predictability. Complex organic systems such as the atmosphere, the human body, the economy, or indeed the universe as a whole are dominated by emergent features as well as interlocking feedback loops, which are difficult to capture using predictive models. One could say that the systems have evolved in such a way that they elude prediction. As discussed also in Chapter 8, simple models which are modest in scope and not restricted by ideas about elegance are usually better at predicting such systems than more complicated models, including those based on rational mechanisms.[49] Clarity and certainty are in short supply.

Living versus non-living. Machines are designed by engineers to be logical and comprehensible. Emergent properties are generally considered to be a bad thing, unless they are of the type which is stable and exact. Living systems, in contrast, are characterized by complex networks, nested feedback loops, and fractal scaling (see Box 10.1). In biology, the only things that are stable are dead (and even they decay).[50]

The emergence of these areas of applied science was assisted by the availability of fast computers starting in the late 1960s, and by a range of new computational tools such as cellular automata and agent-based models. The latter, which became popular in the 1990s, simulate interactions between individual agents: in biology they may represent proteins within a cell or individual cells within an organ; in ecology, members of different species; or in economics, people or firms. The behavior of each agent is determined by a relatively short set of rules. As with cellular automata, running the simulation often results in emergent behavior which is easy to recognize but impossible to predict from a knowledge of the individual agents alone. The models can represent the diversity, dynamism, and collective organization that make the system tick.

Such models have none of the formal aesthetic appeal of the models used in theoretical physics. They rate poorly for harmony and symmetry. Stability is an emergent property that is usually provisional, imperfect, and subject to change. The global properties are not reducible to formal equations or elegant mathematical shortcuts. But that doesn't mean that the results are without aesthetic interest. As the mathematician Jonathan M. Borewein notes, mathematical thought patterns change with time, and "these in turn affect aesthetic criteria—not only in terms of what counts as an interesting problem, but also what methods the mathematician can use to approach these problems, as well as how a mathematician judges their solutions. As mathematics becomes more 'biological,' and more computational, aesthetic criteria will continue to change."[51]

Part of this aesthetic shift is an interest in the emergent properties revealed by figures, rather than just the equations themselves. The most famous examples are fractals such as the Mandelbrot set. This figure, easily viewed on the Internet, is a strange bulbous growth which throws off a universe of organic-looking forms, including infinitely many copies of itself at smaller scales. Despite its immense complexity, it is derived from a very simple equation that is iterated many times for each point in the complex plane. The discovery of chaos in the 1960s revealed that the solutions to many classes of equations were unstable and chaotic but nonetheless had a characteristic shape known as an attractor, which had its own intriguing features.

Here is an example from my own experience: when I worked on magnet design, part of my job was to measure the harmonics of magnets as the current was ramped up or down. The harmonics of a magnet are similar in principle to the harmonics of a note played on an instrument: there is a dominant "tone" which for a bending magnet will be a dipole (with a North pole and a South pole like a bar magnet), plus a mix of other harmonics which are usually error terms that the engineer tries to remove. When I did a similar experiment for a simple set of mathematical equations (using a toy weather model) the result was very surprising.[52]

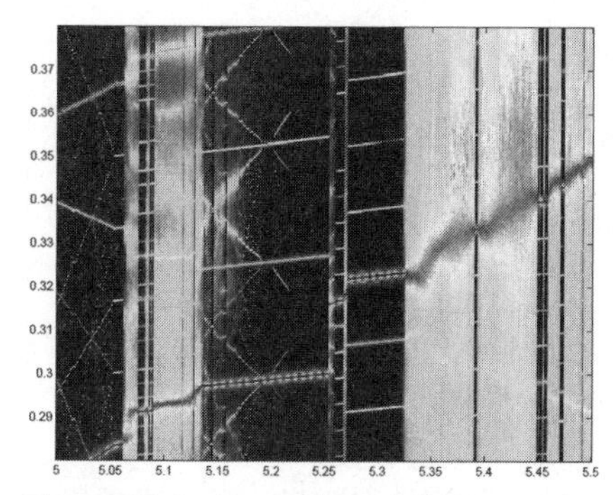

Figure 10.1 Zoomed view of a spectral bifurcation diagram. Horizontal axis is the power level, vertical axis is harmonic frequency, grayscale represents the strength of that frequency at each value of forcing. Source: Author

The model had a single parameter which could be interpreted as a kind of power level for the system. As it was increased, harmonic analysis revealed an intricate, emergent pattern which could never have been predicted from the simple equations themselves (Figure 10.1).

For example, at a power of around 5.3 (on the horizontal scale) there is what resembles a ladder, whose straight rungs correspond to multiples of a base frequency or tone. The dominant tone is the third rung, which is thicker than the others. The relationship between the tones is exact: if you only ran the equations at this power level, you might mistake it for a fundamental property. When the forcing is increased to about 5.32, however, the emergent behavior of the system suddenly switches from exact to a chaotic cacophony of tones. The harmonics of an instrument may be based on Pythagorean ratios, but these appear to have a Hendrix-style fuzz pedal attached. Clearly, such figures are not particularly beautiful, and they do not tell us much about the structure of the universe or the harmony of the spheres; but in aesthetic terms they do change the way we look at something like Figure 7.3, which showed power levels ramping up for the fundamental forces in perfectly straight lines that are assumed to continue forever.

Box 10.1

Fractal Geometry

In *The Assayer* (1623), Galileo wrote that the book of the universe "is written in the mathematical language, and the symbols are triangles, circles and other geometrical figures, without whose help it is impossible to comprehend a single word of it; without which one wanders in vain through a dark labyrinth."[53]

In *The Fractal Geometry of Nature* (1982), however, the mathematician Benoît Mandelbrot pointed out that "clouds are not spheres, mountains are not cones, coastlines are not circles, and bark is not smooth nor does light travel in a straight line."[54] It seems that, in the same way that there are organic and inorganic versions of chemistry, there are organic and inorganic geometries.

Mandelbrot coined the term "fractal," after the Latin *fractus* for broken, to describe this new type of crooked geometry. He was inspired not just by nature, but also by his study of financial data. Financial charts often show a degree of self-similarity, in the sense that plots of price changes look rather alike whether the time scale is a day, week, month, or year. In the same way that clouds have a similar appearance whether viewed up close from an airplane or from far above by a satellite, they are self-similar over different scales.[55]

Price changes also do not vary in a smooth, continuous fashion, but are punctuated by sudden jumps or crashes. Like the coastline of Britain, they are rough and jagged, and the roughness doesn't go away the further you zoom in (as Roger Penrose points out, the idea that cosmic inflation can smooth the universe assumes that the initial structure was not fractal).[56] However, their self-similarity reveals a different kind of order.

A common property of fractal objects is that their features exhibit what is known as scale-free, power-law statistics. There is no typical size or scale: the only rule is that the larger an event or feature is, the less likely it is to happen. For example, if you double the size of an earthquake, it becomes about four times rarer, so earthquake frequency depends on size squared, or size to the power of two—hence "power law." There is no such thing as a "normal" pattern and extreme events are part of the landscape. Similar relationships hold for price changes in a stock market, the size of craters on the Moon, the diameters of blood vessels, the populations of cities, wealth distribution in societies, and many other phenomena.[57]

Many networks found in nature—consisting, for example, of interacting proteins in the human body or species in an ecosystem—also exhibit power-law statistics: a small number of nodes or elements in the network are connected to many others while most nodes have few connections. Human networks such as the World Wide Web or the financial system have a similar pattern, but there is no simple explanation for their appearance—a fact which mathematician Rudy Rucker describes as "elegantly simple, deeply mysterious, but more galling than beautiful."[58] Fractal analysis has revealed a crooked beauty in these systems that has little to do with classical ideas of elegance.

Life Science

Much of the formal development of complexity theory took place in the study of the self-organization of chemical and physical structures by scientists such as the chemist Ilya Prigogine (1917–2003); and in computational fields such as the study of cellular automata. However, the areas where the new techniques and aesthetics seem to be having their greatest impact are in life sciences such as biology and (more recently) economics.

One of the main findings from the Human Genome Project has been that individual genes and proteins rarely act alone; instead, their function can only be understood when they are considered as part of a connected network. Figure 10.2 from a paper by a team of computational biologists (Wang et al. 2009) shows a protein-protein interaction network for

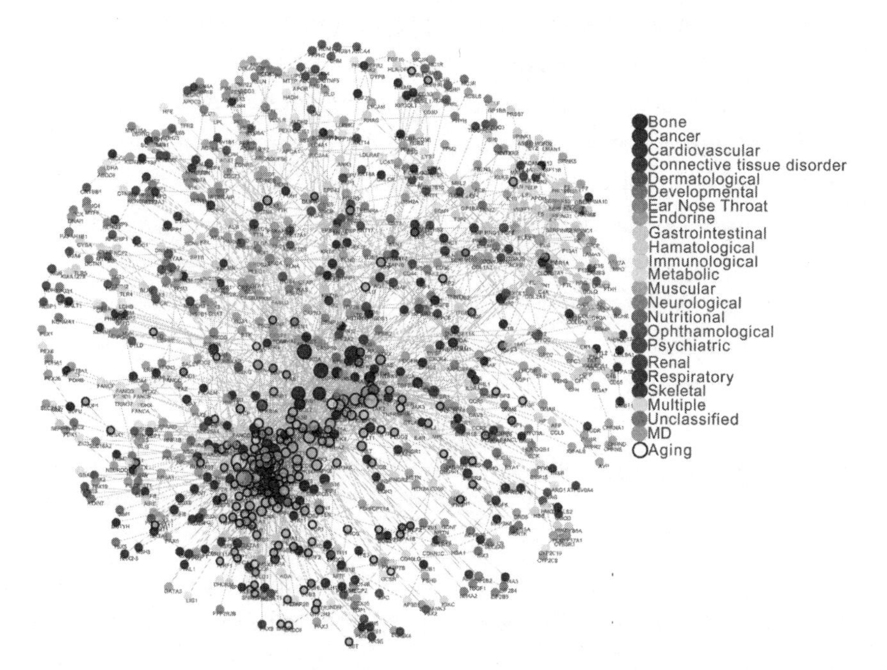

Figure 10.2 Protein-protein interaction network for genes related to different diseases. Nodes represent proteins, lines represent interactions. For details, see Wang et al. 2009. Original version is in color. Source: Reprinted from J. Wang, S. Zhang, Y. Wang, L. Chen, and X-S. Zhang. "Disease-aging network reveals significant roles of aging genes in connecting genetic diseases." *PLoS Comput. Biol.* 5, no. 9 (2009): e1000521.

genes related to a number of diseases. The nodes represent proteins, and lines between nodes represent interactions. Such figures can be used, for example, to understand the relationship between different diseases and suggest appropriate targets for drugs. In this plot the information is displayed as points on a sphere for visual clarity, but in aesthetic terms what is expressed is a sense of multifaceted connection and communication—like peering into the workings of a city's telephone exchange.

Analyzing these complex networks requires expertise not just in biology, but also in chemistry, physics, mathematics, computer science, and so on. This type of integrated science prospers in institutions such as Seattle's Institute for Systems Biology, which have been designed to encourage communication between people from different disciplines rather than atomize them into separate departments.[59]

Scientists are also beginning to apply some of the same systems techniques to the study of the economy. It might seem that economics is inherently a pseudoscience which can tell us nothing interesting about science in general. But I would argue that it is only the most glaring and easily understood example of the perfect model syndrome and our obsession with machine aesthetics. And economic theories have been surprisingly influential in the past. As mentioned in Chapter 8, the mechanism behind Darwin's natural selection was based on Malthusian population dynamics. Neo-Darwinism in turn helped to inspire the multiverse theories of modern cosmology. In fact, the entire Pythagorean project to reduce the universe to numbers was related to the development of money. As the scholar W.K.C. Guthrie observed, it is likely that "Pythagoras derived his enthusiasm for the study of number from its practical applications in commerce," and he was likely also involved in the design of coinage for his region of southern Italy.[60] The famous statement by John Maynard Keynes, "Practical men, who believe themselves to be quite exempt from any intellectual influences, are usually the slaves of some defunct economist," could also apply to scientists.[61]

As the physicist Frederick Soddy pointed out in 1922, economics is located at the nexus between science and spirit: "It is in this middle field that economics lies, unaffected whether by the ultimate philosophy of the electron or the soul, and concerned rather with the *interaction*, with the middle world of life of these two end worlds of physics and mind in their commonest everyday aspects."[62] Following the recent financial crisis, economics is undergoing something of a revolution, similar in spirit at least to the quantum revolution that Soddy and Rutherford helped to create in physics. The hard atom of rational economic man is being replaced by something much more complex, fuzzy, and entangled. We seem finally to be absorbing the lessons from nature that we weren't ready for a century ago. Economics is not much good as a hard science, but it might work as a soft one.

As with quantum physics, the economy is full of right-brain-pleasing paradoxes. For example, Keynes pointed out the paradox of thrift: if one person saves, then that is good; but if everyone saves, then that deprives the economy of funds and leads to recession. There is also a similar paradox related to spending. If one person buys an expensive automobile, it might make them happier in the short term; but it will also make other people more conscious of their own car's relative inferiority. And if everyone does the same thing and upgrades to a better car, then the net gain in happiness soon disappears. As John Stuart Mill put it, "Men do not desire to be rich, but to be richer than other men."[63]

Given the paradoxical nature of the economy—along with the fact that money is a highly emotional subject for most people—it is paradoxical that it has somehow become thought of as rational, efficient, stable, and amenable to the highly linear logic beloved by most economists. Money is a left-hemisphere invention, a means of reducing complexity and plurality to simple numbers. But the modern financial system is something else entirely. As with systems biology, it requires an approach that takes advantage of the massive amounts of data now available, and marries computational and analytical power with a holistic perspective.

Methods from complexity science are now beginning to find widespread use in economics. For example, network theory is used to model how crises are propagated through the financial system and create systemic risk. As the Bank of England's Andrew Haldane notes, risk measurement has traditionally been "atomistic. . . . Risks are evaluated node by node." In a network, however, "this approach gives little sense of risks to the nodes, much less to the overall system."[64] This is especially a concern because of the way in which the world financial system has become increasingly connected. Information propagates rapidly, but so do problems. Robust networks, such as those found in healthy ecosystems, tend to be built up from smaller, weakly connected sub-networks.[65]

A 2011 study by complexity scientists from the Swiss Federal Institute of Technology analyzed the direct and indirect ownership links between 43,000 transnational corporations to make a map of financial power in the global economy (Figure 10.3). The method used was similar to that of the protein-protein network in Figure 10.2. The map showed that "less than 1 per cent of the companies were able to control 40 per cent of

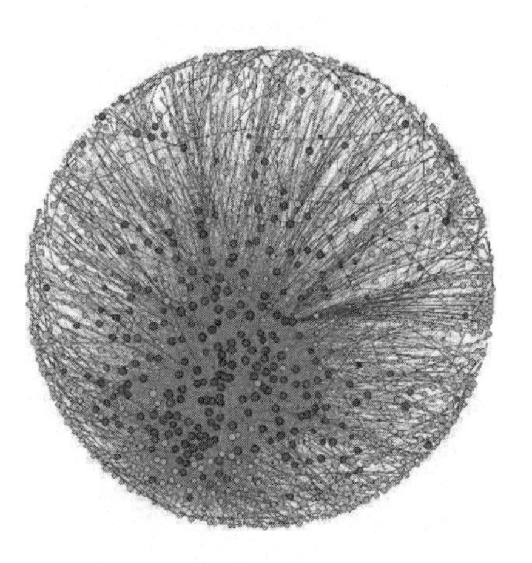

Figure 10.3 Spheres of influence. Upper panel shows network of transnational companies connected through direct and indirect ownership. Lower panel is a zoomed view of some of the most highly connected companies. Nodes represent companies, lines represent ownership links. From Vitali, Glattfelder, and Battiston (2011). Original version is in color. Source: Reprinted from S. Vitali, J.B. Glattfelder, and S. Battiston. "The network of global corporate control." *PLoS ONE* 6, no. 10 (2011): e25995.

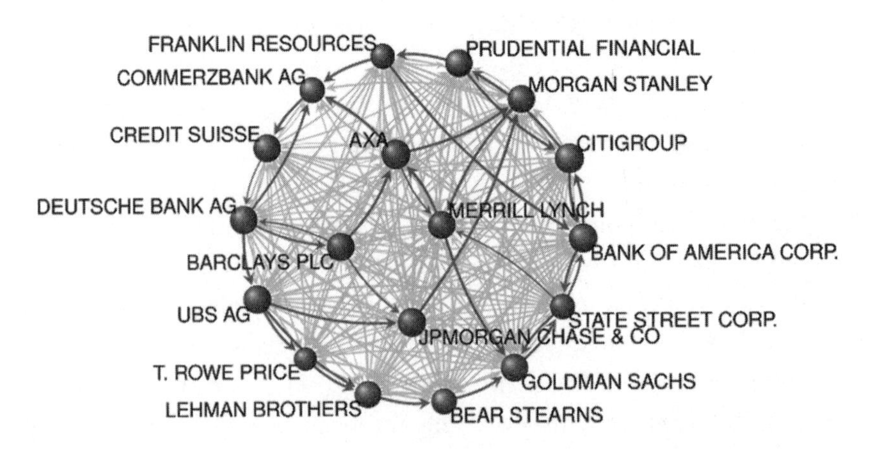

the entire network," according to James Glattfelder, one of the researchers.[66] Most of these powerful companies were financial institutions from the "virtual" economy (i.e., companies that make money out of money), with Barclays Bank, JPMorgan Chase, and Merrill Lynch all in the top ten. The dataset from 2007 also included Lehman Brothers, which was ranked at number 34. Its collapse the next year triggered a global financial crisis and graphically illustrated the system's fragility to the loss of one of these nodes.

As the researchers point out, this dominance by a small group can be viewed as the outcome of a natural process and does not demonstrate conspiracy or collusion. The distribution of power in the economy is related to the fractal structure discussed above, which characterizes many natural systems. But it is clear from the image that the symmetrical neoclassical picture, which sees the economy as being made up of independent "average" firms of similar power, is rather misleading. It is also interesting that we are using the same tools to map the bloated financial sector—which many believe has grown to a size where it has become parasitical to the real economy—as we are to map the network of biological disease processes.

Future Scenarios

As shown by the subprime mortgage crisis of 2007–08, which went almost universally unpredicted, a new approach to economic forecasting would also be useful—and here again are implications for aesthetics and our way of seeing the world. Studies have shown that the left and right hemispheres take quite different approaches to prediction. In one experiment, split-brain patients were shown a sequence of red and green dots that was random but biased, so reds appeared 75 percent of the time and greens only 25 percent. It was found that the right hemisphere tends to guess that the next dot will be red (i.e., takes a simple statistical approach), while the left hemisphere searches for a complicated but nonexistent pattern and does less well.[67] The left hemisphere also tends to be more optimistic than the right, more confident in its own forecasts, and more concerned with rationalizing its forecasts after the fact.[68] Obviously neither approach is perfect, and in most cases it is best to use a combination of techniques. But where the empirical evidence shows that even our most elaborate models do not work, then rather than rationalize their failure using complicated theories we should simply accept that the system is unpredictable and concentrate our attention on finding ways to make it more healthy and stable (in the case of the virtual economy, for example, by better regulating it).[69]

One technique when faced with such uncertainty is to generate a number of scenarios that aim to account for the likely extreme cases and can be used for setting strategy or contingency planning. Scenario forecasting was another spin-off from the Cold War; it was developed when scientists at RAND were attempting to work out what might happen after a nuclear war (i.e., if MAD failed). Its use in business was pioneered by Shell, who credited it with preparing them for the oil price shocks of the 1970s, and it is becoming increasingly widely used. Usually a small number of scenarios, such as two to four, are chosen to represent extreme cases. This helps separate the scenarios from each other (and also accounts for the fact that the future often does turn out to be extreme). As Adam Gordon points out in his book *Future Savvy*, the goal of scenarios is not to make accurate forecasts, but rather to "reach for the storytelling, narrative tradition in human cognition, which is an ancient way of capturing the imagination, educating ourselves and others, and grappling with the complex tapestry of life."[70] Mathematical models can also be used to flesh out the details of the scenarios and check for consistency without masquerading as predictions.[71]

The most important economic question is how our economy will impact the environment, and here the uncertainties—and the need for a pragmatic approach—are even greater. As with climate change, the role of the scientist switches from predicting the system to assessing its health, and even learning from it. Physicists worry themselves about the fine-tuning problem in the Standard Model and ask how it can be explained. But a more urgent question is how we can learn from the fine-tuning that we see in natural systems—which are controlled and regulated by multiple feedback loops—and apply that knowledge to our own behavior. The harmony we need is not with the spheres or the strings, but with the world—and at the moment, we are badly off-pitch.

Box 10.2

The Mind's Eye

The two hemispheres play complementary and sequential roles during perception. Suppose a person is taking a walk out in the forest. The right brain unconsciously scans the overall visual field. When something of interest is detected—say a bird in a tree—the left brain focuses the eyes in on it. The part of the visual field that is brought into sharp resolution is very narrow—only about 3 degrees. The information is then handed back to the right hemisphere, and the end result is a synthesis of right with left. A birdwatcher might take note of the species; a nature enthusiast might enjoy the color of its plumage; a hunter might lift his gun. Normally, the process of perception is fast—the mere flash of an eye. But the human race as a whole seems to be doing something similar, only this time on the scale of millennia.

The philosopher Owen Barfield divided the development of the human consciousness into three phases, which bear close resemblance to these three stages of perception.[72] The first phase, which he called original or unconscious participation, stretched from the Paleolithic to roughly the late Bronze Age or early Iron Age. It was characterized by an instinctive affinity between man and nature, which were seen not as separate but as being made of the same stuff. We were walking in the forest, taking everything in.

The second phase, separation, represented a withdrawal and division from nature. This bifurcation coincided with solarization, when Earth and Moon deities were replaced by solar versions. We stood back from the world so as to be able to understand and manipulate it. We charted the movements of the heavenly bodies and learned to predict them; and we began to use money to facilitate trade. The left hemisphere was turning on and taking control. In Greece, the direction of writing even reversed; people switched to writing from left to right, which is the direction favored by the left hemisphere.[73]

The third stage, which Barfield called final or conscious participation, represents a return to nature—not to our original unconscious relationship, but to one where we can consciously participate in nature through imagination. This corresponds to an integration of the left- and right-brain approaches.

Our perception of the world, seen from a historical perspective, is therefore following a path similar to the cognitive act of perception, but in slow motion. Reductionist science is a particular way of seeing which has been dominated by left-hemisphere functions. We have separated ourselves from nature, named the species of animals and of subatomic particles, and learned to control them. We are now bumping up against both the limits of that control and the limits of nature to support us, and we are seeking a more holistic approach that is in tune with life rather than machines. Areas of science such as complexity theory can be seen as an attempt to reintegrate these different ways of seeing.

The Weak Force

Ideas from complexity theory have also found applications in fundamental theories of physics. In the same way that the properties of matter are collective emergent phenomena, things as basic as time, space, and gravity may also emerge from some deeper structure.[74] In Erik Verlinde's theory of "entropic gravity," for example, gravity is seen as an emergent phenomenon rather than a fundamental force.[75] Since simple programs such as cellular automata are capable of producing complex emergent behavior, some believe that the universe could similarly be generated by just a few concise laws that, when discovered, would form the basis for a new Theory of Everything.[76] However, the fact that simple rules can generate infinite complexity does not imply that the universe can be reduced to simple rules, any more than the fact that simple equations can produce unpredictable behavior means that unpredictable systems can be simulated by simple equations—as many believed after the discovery of chaos. I believe that the message from complexity is different. There is no reason to assume that the universe is simple. That faith—which underlies the quest for a Theory of Everything—is based on an ancient set of aesthetic principles, which applies well to machines but less well to living systems like the universe (which did after all produce us, and includes us).

In the same way that each organizational level of a complex adaptive system shows emergent properties which cannot be deduced from the level below, so a similar phenomenon occurs in physics. "From the emergentist perspective," writes Robert Laughlin, "physical law is a rule of collective behavior, it is a consequence of more primitive rules of behavior underneath (although it need not have been), and it gives one predictive power over a limited range of circumstances. Outside this range, it becomes irrelevant, supplanted by other rules that are either its children or its parents in a hierarchy of descent."[77] Models therefore apply at the level for which they are designed, and the answers we get from interrogating a system depend on the context. This is why the results of accelerator experiments, while interesting and worthwhile, are in strictly practical terms useful mostly for determining what will happen in other accelerator experiments. As with art, their value lies in something other than utility.

It is ironic, then, that science and art have become more polarized than ever. The aesthetics of science have been hidden away behind intimidating equations. Science's exalted, ecstatic ideals of harmony, symmetry, and unity are an intrusion—first initiated by Pythagoras—into the world of art, which has yet to be properly reciprocated. The fascination with beauty and elegance has motivated generations of scientists, and simple solutions to complex problems—such as Einstein's general relativity or the discovery of DNA—will always hold immense power. But the quest for unification can become counterproductive—because it acts as a distraction, and also because its stance is one of control and manipulation. We need to learn from nature, not just exploit it; to see it as a living wonder rather than an inert machine. Science has to reimagine its sense of beauty, but to do that it needs artists as much as it needs great theorists.

The poet Muriel Rukeyser wrote, "The universe is made of stories, not of atoms."[78] One of our most marvelous, enduring, and productive stories, acted out in various guises and in everything from physics to finance, is the story of atoms: with its chance and necessity, its laws and its mutants, its determinism married with passivity, its pared-down elegance and square symmetries, its power and clarity, its *principium individuationis* and *Homo economicus*, its selfish genes longing for eternity, its intense yearning for the One. But there are other stories we can tell. So perhaps the next great science project will be not to find a new particle, or to explain the birth of time, but to hold the poles of science and art together (as with the weak force, this relies on proximity), blow the grid, and forge a new narrative; to teach us how to lock arms and move in rhythm; to go superconducting.

In many ways, science strengthens our appreciation of the natural world through both theory and observation. As Carl Sagan pointed out,

"It does no harm to the romance of the sunset to know a little bit about it."[79] Richard Feynman wrote, "There are all kinds of interesting questions that come from a knowledge of science, which only adds to the excitement and mystery and awe of a flower. It only adds."[80] However, while no one questions that science has opened up new vistas of understanding, our mental models of the world with their crystalline orbs can sometimes get in the way. Science has taught our species a great deal about how the planet works; but our appreciation of it, in economic terms at least, has not grown commensurately, which is one reason why we have an environmental crisis. Models provide a virtual copy of the world. The more perfect it appears, the greater the illusion. We risk seeing the model of the sunset rather than the sunset itself.

We all rely on mental models to make sense of reality and to provide a feeling of stability, and we are willing to expend enormous energy on maintaining them. Science and art are ways to develop, hone, and question these models. It can therefore be disorienting, but also liberating (like throwing away the map and wandering free) to acknowledge their frailty. It reminds us also of the everyday miracle of existence—the fact that everything has emerged out of nothing, against all the odds: you, me, the table I write on, the planet we share. Like the universe itself, our models are impermanent, flawed, and delicately contrived—and that perhaps is the most beautiful thing about them.

Notes

Introduction

1. B. Russell 1898.
2. Hardy 1940.
3. Poincaré 1952, 59–60.
4. Dirac 1963.
5. Chandrasekhar 1997, 65–6; Dyson 1956.
6. Wigner 1960.
7. Gell-Mann 2007.
8. Zee 1986, 3.
9. Wilson 1969.
10. Fry 1926.
11. Scruton 1999, 63–64.
12. Quoted in Sample 2011a.
13. Weinberg 2001, 39.
14. This attribution appeared in an issue of *Reader's Digest* from October 1977. The original source for this comment is probably his statement that "the supreme goal of all theory is to make the irreducible basic elements as simple and as few as possible without having to surrender the adequate representation of a single datum of experience." Einstein 1933, 10.
15. As Anthony Zee notes, "When the beholder is a physicist, beauty *means* symmetry." Zee 1986, 13.
16. Lederman and Teresi 1993, 346.
17. Barrow 2007, 137.
18. Tegmark 2006.
19. Dawkins 1998, x.
20. Koestler 1989, 545.
21. According to Capra, the quest for symmetry "reflects a philosophical attitude which has been inherited from the ancient Greeks and cultivated throughout many centuries." Capra 1983, 257. For critiques of the book, see Woit 2006, 151–52; Lederman and Teresi 1993, 189–91.
22. Lindley 1993, 255.
23. Horgan 1997.
24. Penrose 2005, 1040.
25. Laughlin 2005, xii, 118.
26. Smolin 2006b, 30.
27. Woit 2006, 201–2.
28. Orrell 2007, 319. The book was published in the United States as *The Future of Everything*. In *The Other Side of the Coin*, I expanded on this theme in the context of economics (Orrell 2008). For example Chapter 9 traced the scientific interest in symmetry and showed how it has affected economic thinking. The same argument,

applied more generally to science, was developed in 2009 into a proposal for a book called *The Tragedy of Physics*, which later evolved into this book.

29. Ploeger 2009.
30. Gleiser 2010, 229. In his 2009 book *The Master and his Emissary*, Iain McGilchrist similarly argues that "the universe has no 'profound symmetry'—rather, a profound asymmetry" (McGilchrist 2009, 13). As examples, he mentions Pasteur's discovery of the left-right asymmetry of biological molecules (we discuss this in Chapter 7), and notes that "since then physicists have deduced that asymmetry must have been a condition of the origin of the universe: it was the discrepancy between the amounts of matter and antimatter that enabled the material universe to come into existence at all, and for there to be something rather than nothing. Such unidirectional processes as time and entropy are perhaps examples of that fundamental asymmetry in the world we inhabit." Another example is the left-handedness of neutrinos (Gleiser 2010, 131; Orrell 2008, 266).
31. See e.g. Linde 2012.
32. Weinberg 2001, 97.
33. While ancient models of the cosmos were based on mechanistic arguments, it was not until the seventeenth century that practitioners of what was known as "mechanical philosophy" began to cast off non-mechanistic ideas such as teleology and fully adopt the machine metaphor. See Shapin 1996, 30.
34. Laughlin 2005, 76.
35. Dirac 1934.
36. Bohm 1980.
37. Quoted in Baxter 2004.

1. Harmony

1. Bhikkhu 2003, 108.
2. A key paper on this subject was Alpher, Bethe, and Gamow 1948, which argued that the big bang would explain the proportions of the elements in the early universe. Bethe's name was added as a play on the first three Greek letters (alpha-beta-gamma). Any resemblance to the humor in the TV show *The Big Bang Theory* is coincidental.
3. Burkert 1972, 377.
4. Iamblichus 1918, 58.
5. Burnet 1920.
6. Fludd 1626.
7. Guthrie 1962, 206.
8. Campbell 1991, 75. As the mathematician Ralph Abraham described it, "the goddess submerged into the collective unconscious, while her statues underwent gender change operations." Abraham 2005, 92.
9. According to Iamblichus, the Pythia told a gem engraver, who was seeking business advice, that his wife would give birth to a son "surpassing in beauty and wisdom all that ever lived." This came as something of a surprise, especially because no one—including the wife—knew she was pregnant. Iamblichus 1918, 4.
10. The journalist and philosopher Arthur Koestler wrote of Pythagoras that "his influence on the ideas, and thereby on the destiny, of the human race was probably greater than that of any single man before or after him." Koestler 1989, 25. Not

everyone thinks that this influence has been positive—the biologist Robert Rosen, for example, associated Pythagoras with "a disastrous turn" whose impact "has spread far beyond mathematics." Rosen 2000, 63.

11. Nietzsche 1967.
12. Hillman 1975.
13. The astronomer Fred Hoyle had predicted in 1953 that "once a photograph of the Earth, taken from the outside, is available—once the sheer isolation of the Earth becomes plain—a new idea as powerful as any in history will let loose." Quoted in Cashford 2002, 364. Indeed, the photographer Galen Rowell called the picture of an "Earthrise" by Apollo 8 astronaut William Anders "the most influential environmental photograph ever taken." Rowell 1995.
14. Aristotle 2009, 15.
15. Iamblichus 1918, 44.
16. In Ferguson 2010, 113.
17. In Kumar 2008, 320.
18. In Satter 2012.
19. Plato 2000, 274.
20. In Ferguson 2010, 153.
21. In Peterson 1993, 37.
22. In Ferguson 2010, 193.
23. Vitruvius 1960, 69.
24. Vitruvius 1960, 8.
25. Parallax was first detected by the mathematician and astronomer Friedrich Bessel.
26. Alighieri 1887.
27. Gilby 1951, 236.
28. Joyce 2003, 213.
29. Alberti 1986.
30. Palladio 1965.
31. Paciolo 1509.
32. Weyl 1952.
33. Barrow 2008, 270.
34. Copernicus cites "the consensus of many centuries" behind the dominant view of cosmology at that time, but the Church was largely responsible for enforcing that consensus. Copernicus 1978.
35. Vitruvius 1960, 70.
36. Zöllner and Nathan 2003, 106.
37. Copernicus 1978.
38. Copernicus 1978.
39. In Koestler 1989, 253.
40. The physicist Anthony Zee says something very similar on the role of symmetry in modern physics: "we cannot bring ourselves to imagine Him preferring ellipses to circles and introducing minute variations on the sly." Zee 1986, 213.
41. In Koestler 1989, 393.
42. In Gribbin 2002, 56.
43. Whitehead 1925, 34.
44. Nicolson 1959.
45. Weinberg 2001, 38.

2. Integrity

1. Laertius 1853, 392.
2. Quoted in Eco 1988, 99.
3. Robinson 1968, 202.
4. The infinite series 1/2 + 1/4 + 1/8 + 1/16 + . . . approaches 1 in the limit, so the series can go on forever but the total length remains finite.
5. Iamblichus 1918, 82.
6. McGilchrist 2009, 178.
7. Lucretius 1951, 31.
8. Shakespeare 1925. It is believed that Shakespeare collaborated with another author on this play.
9. Johnson 2008, 49.
10. Lavoisier 1965.
11. In Stewart 2007, 77.
12. Dalton 1834.
13. Dalton 1801.
14. Dalton 1808.
15. Newlands 1865.
16. Mendeleev 1891.
17. Du Fay 1734.
18. Coulomb 1785.
19. Ørsted 1998.
20. Faraday 1855.
21. See the Philosophy Gift Shop, http://www.cafepress.co.uk/philosophy_shop/973063.
22. Maxwell 1865.
23. Hinton 1976.
24. Laughlin 2005, 118.
25. Miller 2006.
26. J.J. Thomson 1897.
27. In Hertz's experiment, residual ionized gas had the effect of shielding the electric field. Thomson managed to achieve a better vacuum and therefore avoid this problem. Jayakumar 2012, 8.
28. J.J. Thomson 1907, 11.
29. J.J. Thomson 1907, 7.
30. J.J. Thomson 1899.
31. Midgley 1985, 114.
32. Bohm 1971, 37.
33. Talbott 2004.
34. Stolnitz 1977.
35. Zee 1986, 3.
36. Tuan 1982, 60.
37. Rand 1938.
38. Nietzsche wrote, "We might call Apollo himself the glorious divine image of the *principium individuationis*, through whose gestures and eyes all the joy and wisdom of 'illusion,' together with its beauty, speak to us." Nietzsche 1967, 36. The term was originally from the work of Arthur Schopenhauer.

39. Harwood Group 1995.
40. Tuan 1982, 139.

3. Radiance

1. J. Palmer 2008.
2. Plato 2000, 175.
3. Plato 1937.
4. From Aristotle's *Politics*. Quoted in Popper 1966.
5. From *The Communist Manifesto*. Quoted in De Botton 2004, 205.
6. Lovejoy 1936, 63.
7. Philoponus 1887.
8. Abraham 2005.
9. V. Galilei 2003.
10. G. Galilei 1914, 178.
11. G. Galilei 1914, 178.
12. G. Galilei 1914, 178.
13. G. Galilei 1914, 179.
14. Newton and Chittenden 1846, 83.
15. One difference is that Galileo equated unimpeded motion on Earth with motion in a circle around the center of the Earth.
16. Descartes 1985, 240.
17. Joule 1845.
18. Von Helmholtz 1853.
19. Noether 1918.
20. Barrow 2007, 63.
21. An early version of some material in Chapters 3 and 5 of this book appears in modified form in Jayakumar 2012.
22. In Quinn 1995, 91.
23. In Bizony 2007, 15.
24. In Kumar 2008, 75.
25. In Lindley 2007, 41.
26. Hannam 2009, 191.
27. In Lederman and Teresi 1993, 153.
28. Quoted in Holton and Brush 2001, 418.
29. Kirkby et al. 2011.
30. The figure shows a pion interaction event, from a 1973 experiment at Fermilab.
31. Rutherford 1911; Rhodes 1986, 50.
32. Quoted in Cathcart 2004, 47.
33. Pauli's proposal was in the form of an open letter addressed to the "Radioactive Ladies and Gentlemen" at a 1930 conference in Tübingen. Pauli 1994, 198–200.
34. Feynman 1964, 4-1.
35. As discussed in the next chapter, it is more accurate to say that mass-energy is conserved, since mass can be transformed into energy or vice versa.
36. Barrow 2007, 63.
37. Plato 1997, 45.

38. Barrow 2007, 211.
39. Zee 1986, 42.
40. Penrose 2005, 22.
41. W. Thomson 1901.

4. The Crooked Universe

1. Miller 2002, 44.
2. As the eighteenth-century philosopher Francis Hutcheson wrote of one proposition in Euclid, it "contains an infinite multitude of truths concerning the infinite possible sizes of right-angled triangles." Hutcheson 1977.
3. Euclid 1908.
4. Saccheri 1920.
5. Einstein in his Kyoto address on 14 December 1922, quoted in Pais and Penrose 2005, 179.
6. G. Galilei 1953, 186–87.
7. Newton and Chittenden 1846, 77.
8. Lecture entitled "Raum und Zeit" (Space and Time), delivered at the annual meeting of the German Association for Natural Scientists and Physicians in Cologne on 21 September 1908; Minkowski 1952.
9. Einstein 1923a.
10. Hafele and Keating 1972.
11. Einstein 1915; Einstein 1916.
12. Letter from Einstein to Heinrich Zangger, 26 November 1915. In Kumar 2008, 123; Einstein 1987, 151.
13. A photograph of the stars was first taken before the eclipse, then compared with one taken during the eclipse. The position of the stars near the sun's edge was predicted to shift by 1.75 seconds of arc, which is about 1,000 times smaller than the angular size of the Sun.
14. In Farmelo 2009, 36.
15. Quoted in Gamow 1970, 44.
16. McIlroy 2011.
17. Aristotle 2009, 213.
18. Planck 1901.
19. Unpublished letter from Max Planck to R.W. Wood, Berlin (1931).
20. Einstein 1905.
21. In Kumar 2008, 51. Millikan was later awarded the Nobel Prize for helping to confirm the theory.
22. Bohr 1913.
23. In Farmelo 2009, 71.
24. In Lindley 2007, 73.
25. In Kumar 2008, 164.
26. Pauli 1925.
27. Fermions were named after Enrico Fermi, bosons after the Indian physicist Satyendranath Bose.
28. Penrose 2005, 100.
29. As discussed later, fermions are made up of three smaller particles known as quarks, while bosons are made from a quark and an antiquark bound together.

30. Einstein 1924.
31. Young 1807.
32. G.I. Taylor 1909.
33. de Broglie 1963, 4.
34. In Mehra and Rechenberg 2000, 604.
35. In Kumar 2008, 206.
36. Schrödinger 1926.
37. In Kumar 2008, 207. The linearity of the wave function is called the Principle of Linear Superposition.
38. In W.J. Moore 1989, 202.
39. Born 1926.
40. Heisenberg 1925.
41. In Kumar 2008, 231.
42. Heisenberg 1958.
43. Dirac 1927; Dirac 1928. Quoted in Webb 2008.
44. In Farmelo 2009, 177.
45. Farmelo 2009, 225.
46. C.D. Anderson 1933.
47. In Farmelo 2009, 336.
48. Heisenberg 1971.
49. Heraclitus 2001, 81:51.
50. Guthrie 1962.
51. Aristotle 2009, 51.
52. In Buckley and Peat 1996 .
53. In Farmelo 2009, 128.
54. Dirac 1930; Farmelo 2009, 178. Like a number of famous scientists including Newton and Einstein, Dirac showed many of the traits of autism. Fitzgerald 2004. In mathematics, there was an even stronger trend towards rigor and aesthetics, exemplified by the Bourbaki group of mathematicians. As the mathematician and economist Roy Weintraub wrote of the Bourbaki movement, "Mathematics was beautiful, and theorems were described in aesthetic terms . . . the highest compliments were of the form 'What a lovely theorem' or 'It's such an elegant proof'." Weintraub 2002, 252.
55. In Kumar 2008, 220.
56. In Kumar 2008, 125.
57. In Kumar 2008, 331.
58. In Trimmer 1980.
59. In Kumar 2008, 356.
60. For Bohm's theory, see Bohm 1980. For a comparison of alternatives, see Penrose 2005, 782–812. Lee Smolin describes the resolution of the foundations of quantum physics as "probably the most serious problem facing modern science." Smolin 2006b, 9. The expression "shut up and calculate" appears to have been first used by David Mermin in a 1989 *Physics Today* article to describe his impression of the Copenhagen interpretation. Mermin 1989. "Most cosmologists . . . reject the Copenhagen interpretation in favor of the quantum multiverse, which describes an infinite ensemble of really existing parallel universes." P. Davies 2007, 257. A poll conducted in July 1999 during a quantum physics conference at Cambridge University found that of ninety physicists polled, four voted for Copenhagen interpretation, thirty favored a modern version of Everett's many worlds, and fifty voted

"none of the above or undecided." Kumar 2008, 258. See also Tegmark 1998. As John Horgan reported from a 1992 symposium on quantum mechanics, "the different interpretations of quantum mechanics cannot be empirically distinguished from one another; philosophers and physicists favor one interpretation over another for aesthetic and philosophical—that is, subjective—reasons." Horgan 1997, 91.

61. Osnaghi, Freitas, and Freire Jr. 2009.
62. According to Heisenberg, "such words as position and velocity and temperature lose their meaning when we get down to the smallest particles." In Buckley and Peat 1996.
63. Quoted in Newton-Smith 2000, 104.
64. In Kumar 2008, 286.
65. In Lindley 2007, 73.
66. In Farmelo 2009, 142.
67. In Kumar 2008, 279.
68. Huxley 1929.
69. Wittgenstein 1980.
70. Miller 2002; Henderson 1983.
71. Durrell 1958.
72. Zee 1986, 77.
73. Braques 1971, 10.
74. Keller 1985, 142.

5. The Masculine Philosophy

1. Steven Weinberg for example argues that "the laws of chemistry arise from the laws of physics" and biological systems can also, in principle at least, be reduced to physics. Weinberg 2001, 212. The "Physics First" movement, championed by Leon Lederman, argues that for related reasons physics should be taught before chemistry and biology. The idea of a hierarchy of sciences was championed in the nineteenth century by the French philosopher August Comte.
2. Horkheimer and Adorno 2002, 4–5.
3. Korsmeyer 2008.
4. Guerrilla Girls 1995.
5. Guthrie 1978, 307.
6. As he wrote in *Politics*, "the male is by nature superior, and the female inferior; and the one rules, and the other is ruled; this principle, of necessity, extends to all mankind." Aristotle 1943.
7. Aquinas 1948.
8. Roszak 1999, 56.
9. See Wertheim 1997 for a discussion of science and gender.
10. Snow and Collini 1993, 103.
11. Wertheim 2006.
12. "Women have been more systematically excluded from doing serious science than from performing any other social activity except, perhaps, frontline warfare. . . . A variety of historical, sociological, and psychological studies explain why this is so, but the fact remains that there are few woman worthies to restore to science's halls of fame. Studies of women's contributions to science have been somewhat more fruitful

though still limited by the same constraints." Harding 1986, 31. A groundbreaking study of the struggle faced by women scientists in America was Rossiter 1982.

13. Roszak 1999, 42.
14. Nochlin 1988.
15. Smolin 2006b, 263. Having worked in both academia and industry, I have noticed that this seems even more true of academia, perhaps to keep up appearances.
16. Smolin 2006b, 336.
17. Randall 2011.
18. Ivie and Tesfaye 2012.
19. Keller 1985, 20. The feminist philosopher and physicist Vandana Shiva—who wrote her 1978 PhD thesis on the philosophical underpinnings of quantum theory—describes the emergence of modern science as "a Western, male-oriented and patriarchal projection which necessarily entailed the subjugation of both nature and women." Shiva 1978, 22.
20. Barres 2006.
21. McCloskey 2000. See also McCloskey 1999.
22. Quoted in Keller 1985, 53.
23. Berger 1972. Nochlin wrote, "There exist, to my knowledge, no historical representations of artists drawing from the nude model which include women in any other role but that of the nude model itself." Nochlin 1988.
24. As feminist theologian and psychologist Catherine Keller notes, the traditional picture of the atom is an entity that "is separate, impenetrable, and only extrinsically and accidentally related to the others it bumps into in its void." Quoted in Roszak 1999, 88. In her anthropological study of particle physicists, Sharon Traweek observes that "Phallic imagery is found in much of the informal discourse of the male particle physicists" (Traweek 1988, 79).
25. Gell-Mann 2007.
26. Zee 1986, 4.
27. Green 2009.
28. Bizony 2007, 81
29. Rutherford 1927.
30. Jayakumar 2012, 37–38.
31. *Reynold's Illustrated News*, 1 May 1932. In Bizony 2007, 87.
32. The details of the reaction are as follows: the polonium emits an alpha particle, which is the same as a helium ion and has an atomic weight of four. This collides with a beryllium nucleus, having an atomic weight of nine, to create a carbon atom of weight twelve, plus a free neutron.
33. Weiner 1969.
34. Frisch 1979, 116.
35. Quoted in Sime 1996, 244.
36. Fermi 1954. Quoted in Gaither and Cavazos-Gaither 2012, 811.
37. Anonymous 1954.
38. Quoted in Bizony 2007, 97.
39. Einstein 1960.
40. Farmelo 2009, 97.
41. See Heisenberg and Heisenberg 2011.
42. Farrell 1945.

43. Laurence 1945.
44. In Davis 1968, 309.
45. Caldicott 1981.
46. Barrow 2008, 492.
47. "Einstein=Man of Conscience^2" 2004.
48. The only scientist to leave the Manhattan Project before its completion was Joseph Rotblat, who with Bertrand Russell went on to found the Pugwash Conferences on Science and World Affairs.
49. "During World War II new kinds of knowledge were identified as necessary for military success. . . . Often the young men (and they were almost all men, not women) who . . . developed these new knowledges, as well as new strategies for making knowledge, themselves became the leaders, at a very young age, of these emerging research fields at American universities. . . . The disciplines most affected were the physical sciences, engineering, economics, political science, psychology, and sociology." Traweek 2000, 38.
50. Finkbeiner 2006, xiv.
51. Defense Advanced Research Projects Agency 2008.
52. Nelson 1996. It is sometimes argued that female scientists work in much the same way as male scientists do. For example, Steven Weinberg writes that "there is no perceptible difference in the way that men and women physicists do physics." Weinberg 2001, 105. He also writes on 152 that "as far as I can see, women and Third World physicists work in just the same way as Western male physicists do." Keller, though, describes the work of Barbara McClintock in genetics, or her own work in biology, as having a fundamentally different—and less aggressively reductionist—character than the mainstream approach. In other areas such as business, it is generally acknowledged that women need to attain a critical mass of 30 percent in order to have a serious influence. Tarr-Whelan 2009.
53. Smolin 2006b, 267.
54. Wertheim 1997, 236.
55. Wertheim 1997, 237.
56. Hawking 1988, 191.
57. Barrow 2007, 112.
58. Gleiser 2010, 125.
59. Farmelo 2009, 74.
60. Einstein, Principles of Research 2009. Address "Motiv des Forschens" to German Physical Society in Berlin, Einstein 1918.
61. As Robert Laughlin wrote, "Nuclear weapons are, unfortunately, the most sensational engineering contribution of physics, something that catapulted the discipline to prominence in the 1950s and has colored it indelibly ever since. This coloring is inherently reductionist." Laughlin 2005, 100.
62. "Theories of deterrence hold axiomatically to the Rational Actor assumption about state behavior and choices with respect to nuclear weapons. Rationality assumes a pure 'cost/benefit' analysis with perfect information that ignores or downplays individual differences among states." Clark 2004, 279.
63. In Abella 2008, 9.
64. Weinberg 2001, 39.
65. Pais and Penrose 2005, 183.

66. Kuhn 1962, 156.
67. Schweber 2011.
68. The National Academies 2012.
69. See Kevles 1995b for a discussion of how postwar science was shaped by a deliberately elitist approach. By 1961, President Eisenhower was warning the country that public policy could "become the captive of a scientific-technological elite." Kevles 1995b, 393.

6. Unity

1. Dirac was rumored to have written a quatrain for a student review: "Age is of course a fever chill/That every physicist must fear/He's better dead than living still/When he's past his thirtieth year."
2. In Holton and Elkana 1982, 225.
3. Isaacson 2007, 466.
4. Holton and Elkana 1982, 224.
5. In Whitrow 1973.
6. Nietzsche 1966, 79.
7. Newton 1730.
8. Zee 1986, 74.
9. Einstein 1923b.
10. "Important though Einstein's formula is," writes Zee, "it is less interesting from an intellectual standpoint than the power of symmetry." Zee 1986, 72.
11. Roughly, F combines the electric and magnetic potentials, while J represents the electric currents which produce those fields. The symbols d and d* are mathematical operators similar to differentiation, but take into account the relativistic curvature of space.
12. Bronowski 1960, 139.
13. Weyl 1918a.
14. Lederman and Teresi 1993, 348.
15. Deutsch 1997, 18
16. P. Davies 2007, 118.
17. Zee 1986, 7
18. Weinberg 1992, 206. For more on Weinberg as the master builder of the Standard Model, see Sutton 2009.
19. Smolin 2006, 18.
20. Barrow 2007, 8
21. Anonymous 1957.
22. Wilson 1987.
23. Lederman and Teresi 1993, 254.
24. Lederman and Teresi 1993, 25–27.
25. Wilson 1987.
26. Ploeger 2009, 4.
27. Lederman and Teresi 1993, 255.
28. Another application is to produce X-rays for microchip lithography. I once worked on the design of a compact superconducting accelerator, about the size of a pool table, which was used for this purpose.
29. In Sample 2010, 127.

30. In Sample 2007.
31. Kevles 1995.
32. Quoted in Kevles 1995.
33. As Ploeger pointed out, the annual budget at Fermilab alone was "staggering when one considers the abstract nature of the work and how difficult it is to explain it to the general public." Ploeger 2009, 11.
34. As Cornell physicist Neil Ashcroft noted, "Condensed-matter physics is at the heart of modern technology, of computer chips, of all the electronic gadgets behind the new industrial order. Yet relative to the big projects, it's neglected." Quoted in Halpern 2009, 159.
35. Einstein 1921.
36. Weinberg 2001, 24.
37. Zee 1986, 273.
38. The person he said this to was a young Leon Lederman. Lederman and Teresi 1993, 15.
39. Chamberlain et al. 1955.
40. In the reaction produced in this experiment, two protons collide to give three protons plus an antiproton.
41. It is possible that right-handed neutrinos exist but for some reason cannot be detected.
42. Quoted in Lederman and Teresi 1993, 256.
43. Danby et al. 1962.
44. Because kaons were produced by strong interactions, physicists believed that they should decay by the same route in an average time of around 10^{-23} seconds (such short-lived particles cannot travel much beyond the atomic nucleus). The explanation, it turned out, was that strange particles decay through the slower weak force.
45. Anonymous 2006a.
46. Gell-Mann 1964.
47. Glashow 1976.
48. The W^- particle acts as an intermediary, so the down quark creates an up quark plus a W^-, and then the W^- decays into the electron and antineutrino.
49. For example, if one quark is ejected from a three-quark baryon, then the energy will create a quark/antiquark pair. One will bond with the ejected quark to form a meson, while the other will bond with the two remaining quarks to form a baryon.
50. Voltaire 1947.
51. Feynman 1948.
52. Quoted in Barrow 2007, 114.
53. Lederman and Teresi 1993, 363.
54. In Calder 2003, 527.
55. Gleiser 2010, 126.
56. Laughlin 2005, 113.
57. Alexander 1920.
58. Laughlin 2005, 34. According to Laughlin, even Newton's laws "are not fundamental at all but a consequence of the aggregation of quantum matter into macroscopic fluids and solids—a collective organizational phenomenon." Laughlin 2005, 31.
59. Laughlin 2005, 89. Interpreting these emergent behaviors has resulted in "a confrontation between reductionist and emergent principles that continues today." Laughlin 2005, 97. See also P.W. Anderson 1972.

7. Broken Mirrors

1. Bacon 1856.
2. Doward and Wilkinson 2011.
3. McGilchrist 2009, 343.
4. Mann 1996, 471. The novel alludes to, and was influenced by, Nietzsche's *Birth of Tragedy*.
5. As Dahlia W. Zaidel and Choi Deblieck write: "it is quite common to see asymmetric moles (e.g., Marilyn Monroe, Cindy Crawford) . . . which do not seem to detract from general facial attractiveness or popularity." Zaidel and Deblieck 2007.
6. J. Palmer 2008.
7. In Peterson 1993, 37.
8. Worthington 1895.
9. Weyl 1918b.
10. As Weinberg put it in his Nobel lecture, "I was impressed by the fact that quantum electrodynamics could in a sense be *derived* from symmetry principles and the constraints of renormalizability; the only Lorentz invariant and gauge invariant renormalizable Lagrangian for photons and electrons is precisely the original Dirac Lagrangian of QED." Weinberg 1980.
11. Glashow 1961.
12. In mathematical terms, we say that the SU(2) group is non-commutative. Because SU(2) uses complex numbers, it has an additional wrinkle: as with a spin-1/2 particle like an electron, in order to get back to the same place you have to rotate not by 360 degrees, but by 720 degrees. So SU(2) can be thought of as a mathematical object representing the symmetries of a strange kind of sphere. Technically, it is known as a "double cover" of the group of rotations in three dimensions.
13. The interaction of the force-mediating particles is related to the non-commutativity of SU(2); see previous note.
14. As Anthony Zee wrote, "It's a description of the world in repose." Zee 1986, 218.
15. Higgs 1964.
16. Lederman and Teresi 1993.
17. Sample 2010, 164.
18. Woit 2006, 100.
19. Glashow 1961; Salam 1968; Weinberg 1967.
20. Weinberg 2001, 186.
21. In technical terms, SU(3) can be described as a 3-sphere bundle over a 5-sphere, where a 3-sphere is a sphere in three dimensions, and a 5-sphere is a sphere in five dimensions. Aguilar and Socolovsky 1999.
22. Gross and Wilczek 1973.
23. Georgi and Glashow 1974. In general, SU(n) requires n^2-1 exchange particles.
24. As Anthony Zee wrote, "Many physicists, myself included, are willing to believe in the Georgi-Glashow theory on aesthetic grounds alone." Zee 1986, 235.
25. In Smolin 2006b, 65.
26. Amaldi, der Boer, and Furstenau 1991. Their results, which incorporated the latest experimental results from LEP, built on earlier work such as Dimopoulos, Raby, and Wilczek 1981.
27. Barrow 2008, 512.
28. Gell-Mann 2007.

29. McMaster 2003b.
30. The anonymous writer continues: "I can't believe some people still reject this and the other hints as evidence that [supersymmetric theory] is the real deal." See http://backreaction.blogspot.com/2007/12/running-coupling-constants.html.
31. In Brumfiel 2011.
32. Gross 2005.
33. Quoted in Castelvecchi 2012.
34. Veneziano 1968. For a general audience discussion, see Peat 1991, 59.
35. The Calabi-Yau manifold was developed by the mathematicians Eugenio Calabi and Shing-tung Yau, and it turned out to be exactly what superstring theory needed. Candelas et al. 1985.
36. Scherk and Schwarz 1974.
37. Green and Schwarz 1984.
38. Anonymous 2001.
39. Anonymous 2007. For a critique of the E8 theory, see Distler and Garibaldi 2010.
40. Lisi and Weatherall 2010.
41. Zetter 2008.
42. Rovelli and Smolin 1990; Smolin 2003. An advantage of the theory, as Smolin points out, is that it is "background independent" in the sense that the relativistic properties of spacetime emerge naturally. See Smolin 2006a.
43. Quoted in Smolin 2006b, 125.
44. McMaster 2003a.
45. Laughlin 2005, 212.
46. Woit 2006.
47. Weinberg 2001, 100.
48. Quoted in Woit 2006, 155.
49. Quoted in Woit 2006, 155.
50. As Peter Woit points out, the beauty of string theory is in large part "the beauty of mystery and magic." Woit 2006, 202.
51. As Tomáš Sedláček notes of the scientific world, "Those who 'create' the truth and those who 'appraise' it are one and the same. . . . there is none of the division of power we know and carefully watch over in the world of politics." The same might be said of those who create and appraise scientific beauty. Sedláček 2011, 239.
52. Matthews 2006.
53. Smolin 2006b, 284.
54. Image by John Stembridge, used with permission. See http://www.aimath.org/E8/mcmullen.html.
55. Fludd 1626.
56. For example, Anthony Zee wrote that "Green and Schwarz discovered that superstring theory possesses some amazingly attractive properties. In particular, thanks to its intricate symmetry structure, the quantum version of superstring theory is renormalizable." Zee 1986, 272.
57. Smolin 2006b, 281.
58. Portuguese cosmologist João Magueijo wrote that, "As with every cult, people who do not conform to the party line are ostracized and persecuted." Magueijo 2003, 236.
59. In Smolin 2006b, 47.
60. "It is anomalous to replace the four-dimensional continuum by a five-dimensional

one and then subsequently to tie up artificially one of those five dimensions in order to account for the fact that it does not manifest itself." Einstein quoted in Smolin 2006b, 48.

61. See Duff and Sutton 1988; Witten 1995. M-theory needs an extra dimension, so in its case there are ten spatial dimensions plus time.
62. According to Lee Smolin, when Witten presented M-theory at a string theory conference in 1995, he didn't say what the M stood for "because the theory did not yet exist. We were invited to fill in the rest of the name by inventing the theory itself." Smolin 2006b, 136. Witten later suggested magic, mystery, or membrane, while Sheldon Glashow joked that it could be an upside-down W. Woit 2006, 163.
63. While 10^{500} is the number most commonly bandied about, the actual number could be infinite. See Douglas 2003; DeWolfe et al. 2005.
64. Susskind 2006. See also Smolin 1992.
65. One such physicist is Stephen Hawking, who wrote that M-theory is "the *only* candidate for a complete theory of the universe." Hawking and Mlodinow 2010, 181.
66. João Magueijo refers to Witten as the "cult leader" of string theory. Magueijo 2003, 239.
67. Lemonick 2004; McMaster 2003b.
68. Woit 2006, 115.
69. McMaster 2003b.
70. Barrow 2007, 216.
71. Penrose 2005, 449.
72. In McGilchrist 2009, 13. McGilchrist notes that other examples of this asymmetry include the imbalance between matter and antimatter, and unidirectional processes such as time and entropy.
73. Koren 1994.
74. "String theory has the remarkable property of predicting gravity." Witten 1996.
75. In Ghosh 2011.
76. In Ghosh 2011.
77. Overbye 2012.
78. See Cho 2007. As Steven Weinberg wrote, "The discovery of the Higgs boson would be a gratifying verification of present theory, but it will not point the way to a more comprehensive future theory. We can hope . . . that the most exciting thing to be discovered at the LHC will be something quite unexpected." Weinberg 2012.
79. For further updates and discussion on the Higgs, supersymmetric particles, etc. see http://futureofeverything.wordpress.com/category/physics-2.
80. See http://www.nationalgallery.org.uk/leonardo-da-vinci-painter-at-the-court-of-milan-exhibition-guide.
81. Amis 1995.

8. The Shadow World

1. Cornford 1922.
2. Plato 1930–35.
3. Kepler 1604.
4. Pope 2006.
5. For a popular account of dark matter and dark energy, see Panek 2011.

6. In Johnson 2008, 49.
7. In Brownlee 2011.
8. Milgrom proposed a scheme known as Modified Newtonian dynamics, or MOND. Milgrom 1983. Alternative formulations of gravity such as MOND have their own problems; however this of course does not imply that the current theory is correct.
9. Smolin 2006b, 15.
10. Motl 2004.
11. In Atkinson 2011.
12. Weinberg 2001, 59.
13. According to Weinberg, "no one doubts that with a large enough computer we could in principle explain all the properties of DNA by solving the equations of quantum mechanics for electrons and the nuclei of a few common elements, whose properties are explained in turn by the standard model." Weinberg 1992, 32. As will be discussed later, plenty of people contest this reductionist paradigm.
14. Vitruvius 1960, 39.
15. The material in the rest of this chapter is based on some talks I gave in 2010 and 2011 called "Perfect Model: Prediction, Science and Aesthetics."
16. The English physicist Lewis Fry Richardson worked out the basic equations during World War I. Richardson 1922.
17. P.N. Edwards 2000.
18. Lorenz 1963. The quote, and the term "butterfly effect," came from a 1972 talk to the Association for the Advancement of Science in Washington, DC. Lorenz 2000.
19. As Lorenz put it, "now we had an excuse." Quoted in Gleick 1987, 18.
20. Lewis 2005.
21. Woods 2005, 118.
22. For example, Toth et al. (1996) wrote that "we will assume that our numerical model is essentially perfect."
23. As Robert Laughlin notes, "The laws of hydrodynamics amount to a precise mathematical codification of the things we intuitively associate with the fluid state. . . . No one has ever succeeded in deducing these laws from first principles, although it is possible to make highly plausible arguments in many cases. The reason we believe them, as with most emergent things, is because we observe them." Laughlin 2005, 40.
24. Wolfram 2002, 381.
25. Orrell 2001; Orrell et al. 2001; Orrell 2002. For popular accounts, see Orrell 2007; Matthews 2001; Calder 2003, 138.
26. Based on my own experience, some meteorologists seem to think that the butterfly effect was proven by Lorenz, while others distance themselves from the "hypothetical butterfly scenario of popular imagination" to an extent (although, of course, it was a product of the scientific imagination—when I attended ensemble forecasting meetings during my D.Phil. research, nearly every presentation seemed to begin with a picture of a butterfly). Wolchover 2011. Paul Roebber, a mathematician and meteorologist at the University of Wisconsin-Milwaukee, told science journalist Natalie Wolchover that "I agree . . . that butterfly-scale effects would get damped out, but influences that are still small-scale influences from a weather perspective, such as individual clouds—those effects are much more likely to grow and be important." As in life, small changes can sometimes have big consequences, but

what counts here is the effect of typical perturbations. Analysis shows that while initial errors are significant at first, they are soon dominated by model error, and chaos is a secondary effect. The reasons for the historical emphasis on sensitivity to initial condition are institutional, not scientific. See Orrell 2007.

27. W. Thomson 1901.
28. Roebber et al. 2004.
29. In Mullin 1989.
30. For a discussion of uncertainty in environmental forecasting, see Beven 2009.
31. Gardner 1970.
32. Wolfram 2002.
33. Gray 2010.
34. Caroe 2011; Anonymous 2011b; Monbiot 2012.
35. For example, it is sometimes argued that a number of models give similar results, so it is unlikely that they are all wrong (Diacu 2009, 174). However, this ignores the tendency of forecasters to share techniques and cluster their predictions. Using the same argument, the fact that none of the forecasters polled by Bloomberg.com at the start of 2008 foresaw a loss for the year, and that the average prediction was for a gain of 11 percent, was convincing proof that the world economy was in good shape (despite that by year-end the S&P 500 index was down 38 percent). Thomasson 2009. Another argument that is sometimes advanced is that climate scientists are only trying to predict the effect of a perturbation caused by carbon emissions to the climate. But this type of prediction is often even harder because the effect of the perturbation is unclear. It is hard enough to predict the effect of a drug on a cancer cell, let alone carbon emissions on the planet.
36. Some who were present at the meeting described the estimate as hand waving. Kerr 2004.
37. The word "canonical" is used in this way in Kerr 2005. Of course, not everyone gets the same results. One 2011 study found a range of 1.7 to 2.6 kelvin with a 66 percent probability after calibrating against conditions 20,000 years ago, but again many assumptions are made about the validity of the model. Schmittner et al. 2011. Also, because the analysis did not account for potential changes in clouds, the lead author admits that "the range that we estimate for climate sensitivity may be too narrow." Nuwer 2011.
38. IPCC 2007.
39. S.H. Schneider 2005.
40. Stainforth et al. 2005. Six parameters were perturbed, which were "the threshold of relative humidity for cloud formation, the cloud-to-rain conversion threshold, the cloud-to-rain conversion rate, the ice fall speed, the cloud fraction at saturation and the convection entrainment rate coefficient." The paper points out that the range of perturbations may be too low because "experts are known to underestimate uncertainty even in straightforward elicitation exercises." Some simulations were omitted because they went unstable or even showed cooling.
41. An example of abrupt climate change was the Younger Dryas period, which began about 12,700 years ago. The warming Earth suddenly plunged back into ice-age conditions and remained cold for over a thousand years. It may have been triggered by an abrupt reversal of the ocean circulation pattern which drives the Gulf Stream.
42. In Wohlforth 2004, 169.

43. In climate change prediction, "simple forecasting methods apparently provide forecasts which are at least as accurate as the much more complex GCMs for forecasting the global temperature." Fildes and Kourentzes 2011. As mentioned in Chapter 10, the same phenomenon is seen in other areas such as economics.
44. As discussed in Orrell 2007, when I began my D.Phil. there had been remarkably little investigation into model error. For another example from a different context, see Keepin and Wynne 1984.
45. Fildes and Kourentzes 2011. As the authors note, while there have been many studies of climate models, "few, if any, studies have made a formal examination of their comparative forecasting accuracy records, which is at the heart of forecasting research" (973). The climate model tested in this study was the UK Met Office's decadal climate prediction system, known as DePreSys. The time-series model, which led to an improved forecast when combined with DePreSys, was a Holt linear trend model. For discussion, see Keenlyside 2011; McSharry 2011.
46. These models are characterized by "a myriad of 'physically meaningful' and interconnected subsystems, each governed by the 'laws of nature,' applied at the microscale but allowed to define the dynamic behaviour at the macroscale, in a manner almost totally specified by the scientist's (and usually his/her peer group's) perception of the system." The idea that models are based on physical laws is also commonly evoked as a defense of their predictive ability. Shackley et al. 1998.
47. See, for example, Coumou and Rahmstorf 2012.
48. For further discussion of this issue, see Stainforth et al. 2007.
49. See http://www.capefarewell.com.
50. Nielsen 2011.
51. Harris 2011.
52. As Fildes and Kourentzes note, "The scientific community of global climate modellers has surely taken unnecessary risks in raising the stakes so high when depending on forecasts and models that have many weaknesses." Fildes and Kourentzes 2011, 992. Forecasting experts Spyros Makridakis and Nassim Taleb added that "it must be accepted that accurate predictions are not possible and uncertainty cannot be reduced (a fact made obvious by the many and contradictory predictions concerning global warming), and whatever actions are taken to protect the environment must be justified based on other reasons than the accurate forecasting of future temperatures." Makridakis and Taleb 2009.
53. Orrell 2006, 311. As Marcelo Gleiser wrote, "The uncertainty of knowledge is as permanent as quantum uncertainty. Hard as this may be to accept, it is a fundamental limitation of human understanding." Gleiser 2010, 98.
54. In Gribbin 2002, 56.
55. Sullivan, Gilbert, and Treharne, 1940.
56. Whitehead 1925, 97.
57. Genes could be either dominant or recessive. For example, suppose that the "smooth" gene is dominant. If a child plant inherits either a smooth gene from one of its parents or smooth genes from each of them, then it will have smooth seeds. It will only have wrinkly seeds if it inherits a wrinkly gene from each parent, which has a probability of one in four assuming that the two genes are equally distributed in the parents. Therefore three quarters of the peas should be smooth and one quarter wrinkly, which was in almost exact agreement with Mendel's experimental results.

58. Morgan et al. 1915.
59. Schrödinger 1944.
60. In Farmelo 2009, 280.
61. Bernal 1958.
62. Watson and Crick 1953.
63. Monod 1971, 112–13.
64. In Walsh 2009.
65. Sample 2011b; European Society of Human Genetics 2011. See also Janssens et al. 2008.
66. Kollewe 2011.
67. Watson 1968, 50.
68. Anker and Nelkin 2004, xii.
69. Anker and Nelkin 2004, 28.
70. Dawkins 1976.
71. Deutsch 1997, 175.
72. See Mattick 2005.
73. For a popular account, see Carey 2011.
74. Tokyo International Forum 2008.
75. Clapworthy et al. 2008.
76. Krause 2000.
77. This is not to say that more specialized models are not commercially viable. One of the leaders in predictive biosimulation is Entelos, who market a Virtual Mouse to reduce the need for animal experiments and Virtual Patients to forecast the results of human drug trials. In 2011 the firm went into Chapter 11 bankruptcy, but emerged later that year. Gartner 2005.
78. See for example Curtis et al. 2012. Disclaimer: I am on the scientific board of a company, Physiomics, which has developed a model—naturally called the Virtual Tumor—to test and optimize the effectiveness of cancer drugs. The model takes an agent-based approach to simulate the effect of anti-cancer drugs on a population of cells. Its predictive ability has been extensively validated against experimental data in single-blind tests. Snell, et al. 2011. See also www.physiomics-plc.com.
79. Tedd Kaptchuk, who one expert called "the most knowledgeable person in the world on all matters placebo," told the *New Yorker* that "I am sure that I do not understand the placebo effect." Specter 2011. It is possible that there are genetic factors behind a person's susceptibility to the placebo effect.
80. Kaptchuk, Kerr, and Zanger 2009; Mesmer 1948.
81. For example, sleeping pills are usually blue because that color enhances the placebo effect (except for Italian males, who associate blue with their national football team). Moerman 2002, 49.
82. In Specter 2011.
83. Leibniz 1902.
84. The theory of utility was founded by the English philosopher Jeremy Bentham, who defined it simplistically as a linear sum of positive pleasures and negative pains. Bentham 1907. According to one study, it is likely that today Bentham would be diagnosed with Asperger's syndrome, and indeed his idea of happiness does seem more concerned with logic than emotion. Lucas and Sheeran 2006.
85. Jevons 1957, xvii–xviii.

86. Quetelet 1842, 100.
87. Edgeworth 1881, 16.
88. Nietzsche's work was not embraced by economists, perhaps because his remarks about them were "almost universally derogatory." Senn 2008.
89. According to Smith, this process was the happy result of our egoistic quest for profit. Smith 1869. As Czech economist Tomáš Sedláček points out, the idea of the invisible hand is much older than Adam Smith and has roots that go back to antiquity. Sedláček 2011, 38.
90. Heims 1980, 295.
91. Arrow and Debreu 1954.
92. This is why hedge funds don't use equilibrium models. McCauley 2004, 168; Makridakis, Hogarth, and Gaba 2009. For plots of forecast error from economic models, see Orrell 2010, 34; Orrell 2007, 242. For a TV documentary on the subject, see *The Prediction Business* 2009.
93. This quotation is from Janet Yellen, then president of the Federal Reserve Bank of San Francisco. Timothy Geithner, who was then president of the Federal Reserve Bank of New York, said, "We think the fundamentals of the expansion going forward still look good." There was also general acclaim for Alan Greenspan, whose greatness Geithner described as not yet fully appreciated. Appelbaum 2012. As late as April 2007 the International Monetary Fund said, "Notwithstanding the recent bout of financial volatility, the world economy still looks well set for continued robust growth in 2007 and 2008." Makridakis, Hogarth, and Gaba 2009. Nassim Taleb wrote, "It appears scandalous that, of the hundreds of thousands of professionals involved, including prime public institutions such as the World Bank, the International Monetary Fund, different governmental agencies and central banks, private institutions such as banks, insurance companies, and large corporations, and, finally, academic departments, only a few individuals considered the possibility of the total collapse of the banking system that started in 2007." Taleb 2009. Greenspan later put the blame for the crisis on an "intellectual edifice" created by Nobel prize-winning economists. Greenspan 2008.
94. Fama 1965.
95. Anonymous 2006b.
96. Buffet 1989.
97. Actually, the butterfly effect has been used as well. In a speech by Federal Reserve Chairman Ben Bernanke, he explained why even the most elaborate economic models consistently get it wrong: "in a sufficiently complex system, a small cause—the flapping of a butterfly's wings in Brazil—might conceivably have a disproportionately large effect—a typhoon in the Pacific." Or a world recession. Bernanke 2009.
98. Bachelier (1900) argued that financial markets follow a random walk. Lo and MacKinlay (1988) later performed a simple statistical test which refuted the claim.
99. Triana 2009.
100. Dawkins 1976, 4.
101. James 2008, 193.
102. "In October 1838, that is, fifteen months after I had begun my systematic inquiry," wrote Charles Darwin in his autobiography, "I happened to read for amusement Malthus on Population, and being well prepared to appreciate the struggle for existence which everywhere goes on from long-continued observation of the habits of

animals and plants, it at once struck me that under these circumstances favourable variations would tend to be preserved, and unfavourable ones to be destroyed. The results of this would be the formation of a new species. Here, then I had at last got a theory by which to work." Darwin 1887; Malthus 1970.

103. As Spencer wrote in his 1851 *Social Statics*, "It seems hard that widows and orphans should be left to struggle for life or death. Nevertheless, when regarded not separately, but in connection with the interests of a universal humanity, these harsh fatalities are seen to be full of the highest beneficence—the same beneficence which brings to early graves the children of diseased parents." Spencer 1851.
104. Anonymous 2011a.
105. Bakan 2004.
106. For a comparison of GDP and reported happiness levels in the United States, see Orrell 2008, 73.
107. Blake 1794.
108. Hannam 2009, 174.
109. A random sample of such complaints: replying to a 1901 letter from the economist Léon Walras, the French mathematician Henri Poincaré warned that the theory's unrealistic assumptions might make the conclusions "devoid of all interest." Beinhocker 2006, 49. The mathematician Norbert Wiener wrote in 1964 that "economists have developed the habit of dressing up their rather imprecise ideas in the language of the infinitesimal calculus. . . . To assign what purports to be precise values to such essentially vague quantities is neither useful nor honest, and any pretense of applying formulae to these loosely defined quantities is a sham and a waste of time." Wiener 1964, 89. Richard Feynman said of the tendency for social scientists to mathematize their work, "I have a great suspicion that they don't know, that this stuff is [wrong] and they're intimidating people." Feynman 2000, 22–23. Systems scientist Jay W. Forrester observed that mainstream economics retains a conceptual framework that "is narrow, is based on unrealistic assumptions, emphasizes equilibrium conditions, and is committed to mostly linear mathematical methods." Forrester 2003. French physicist and hedge fund manager Jean-Philippe Bouchaud wrote, "We need to break away from classical economics and develop completely different tools." Bouchaud 2008. Economist Paul Krugman said in a 2009 speech that much of the past thirty years of macroeconomics was "spectacularly useless at best, and positively harmful at worst." Anonymous 2009.
110. Veblen 1898.
111. This field was started by the psychologists Daniel Kahneman and Amos Tversky in the 1970s. See Kahneman and Tversky 1979; Kahneman 2011.
112. Bushong et al. 2010.
113. See, for example, McClure et al. 2004.
114. As Eugene Fama notes, "Tests of market efficiency are tests of some model of market equilibrium and vice versa. The two are joined at the hip." In Clement 2007.
115. For a discussion of systems that are far from equilibrium, see Bak 1996. As Jean-Philippe Bouchaud notes, mainstream economics "has no framework through which to understand 'wild' markets, even though their existence is so obvious to the layman." Bouchaud 2008. See also Chapter 10, note 57.
116. Bateson 1999.
117. Bowles 2009.

118. McCloskey and Ziliak 2001, 303.
119. Häring and Douglas 2012, x. Economists M. Neil Browne and J. Kevin Quinn also describe the "almost complete absence of power from the toolkit employed by mainstream economists." Browne and Quinn 2008.
120. Sedláček 2011, 183.
121. Soddy 1926. Soddy was derided as a "crank" for his thinking about monetary policy, even though many of his recommendations, such as switching from the gold standard to floating exchange rates, eventually became orthodoxy. A.S. Russell 1956.
122. As an article in the Economist noted, "If asset prices reflect economic fundamentals, why not just model the fundamentals, ignoring the shadow they cast on Wall Street?" Specialized models for studying finance do exist, but they are considered "very difficult to handle," which means that things like bank runs don't get much attention. Anonymous 2009. One of the few economists to investigate the instability of the financial system was Hyman Minsky; see Minsky 1972.
123. See, for example, Daly 1996.
124. Noble 2006.
125. Friedman 1953.
126. Krugman 2009.
127. In H. Schneider 2011.
128. See for example Klein 2007; Lasn and Adbusters 2012; Orrell 2008; Orrell 2010; Sedláček, Orrell, and Chlupatý 2012.
129. According to a United Nations report, in 2000 the richest 2 percent of the world population owned more than half of the world's financial assets. Davies et al. 2006.
130. Hirst's "For the Love of God" is a platinum cast of a human skull encrusted with 8,601 diamonds, reputedly sold in 2007 for an asking price of £50 million. Many of Hirst's works, such as his series of dot paintings, are produced not by him but by his assistants, thus blurring the line between genuine and fake. The art market is driven by the richest portion of society and according to one study is an indicator for social inequality: "a one percentage point increase in the share of total income earned by the top 0.1% triggers an increase in art prices of about 14 percent." Goetzmann, Renneboog, and Spaenjers 2009. The art market is basically a branch of the luxury goods business; see Fraser 2011.
131. Soddy 2003.
132. Weinberg 2001, 106.
133. Quoted in Salam 1990.
134. Smolin 2003.
135. In Wroe 2012.
136. Levins 1966.
137. Allen 1971.

9. The Virtual Universe

1. In Metzner 1998, 163.
2. In Metzner 1998, 165.
3. Vaucanson 1985.
4. In Wood 2002.
5. Quoted in Capra 1996, 9.
6. Quoted in Brummett 1999, 15.

7. Quoted in Brummett 1999, 15.
8. Quoted in Brummett 1999, 15.
9. Dawkins 1991.
10. Whitehead 1920.
11. Discussed in Weinberg 1992, 255.
12. Campbell 1995.
13. Wesson 1994, 308.
14. In Kroehling 1991.
15. Le Corbusier 1971.
16. A picture of Le Corbusier's design for such a city can be found at http://www.athenaeum.ch/corbu3m1.htm.
17. M.C. Taylor 2001.
18. R. Moore 2012.
19. Everett 1957.
20. Osnaghi, Freitas, and Freire Jr. 2009.
21. See Chapter 4, note 60.
22. Deutsch 1997, 51.
23. Tipler 2012.
24. Davies 2007, 147.
25. Guth 1981; Sato 1981.
26. Penrose 2005, 754. As Robert Laughlin points out: "Theories of explosions, including the first picoseconds of the big bang . . . are inherently unfalsifiable, notwithstanding widely cited supporting 'evidence' such as isotopic abundances at the surface of stars and the cosmic microwave background anisotropy. One might as well claim to infer the properties of atoms from the storm damage of a hurricane." Laughlin 2005, 211.
27. Hoyle 1982.
28. Chaotic or eternal inflation was first proposed by Andrei Linde (1986). Practically all models of inflation predict eternal inflation. See also Adler 2011.
29. Tegmark 2006. Tegmark's theory bears some similarity to the "modal realism" of philosopher David Kellogg Lewis: "the thesis that the world we are part of is but one of a plurality of worlds." D.K. Lewis 1986, vii.
30. Dirac 1963.
31. Linde 2012.
32. Dawkins 2006, 173.
33. Dawkins 2006, 175–76.
34. Dawkins 2006, 166.
35. Brian Greene described multiverse theory as "high-risk science" but surely the opposite is true, since it can never be falsified. Greene 2012.
36. In Davies 2007, 179.
37. The complexity researcher Stephen Wolfram, for example, notes that cellular automata are capable of producing emergent behavior that is uncomputably complex, despite the fact that the local rules used to generate it are very simple. If space is arranged in a discrete fashion, then some kind of similar rules could generate all the complexity that we see in the universe. As he told *Wired* magazine in 2002, the rule might extend to only "three, four lines of code." Levy 2002.
38. Descartes 1988, 85.
39. Curtis 2011.

40. Part of the appeal of the Occupy street protests held around the world in 2011 was the way in which they exerted a real physical presence in places like Wall Street, and therefore reclaimed the economy from the abstract, virtual machinations of financiers.
41. Anonymous 2011c.
42. Of course, there have been occasions when theory has been ahead of observations, as when Dirac predicted antimatter. But the extrapolations being made here are of a different scale altogether.
43. Hannam 2009, 103.
44. Adams 1985.
45. Wertheim 2011, 266. Far from being an outsider event, the conference was attended by luminaries including Brian Greene (who was busy doing interviews for the PBS series *The Elegant Universe*), Stephen Hawking, and Lisa Randall.
46. Von Franz 1968, 234.
47. See Davies 2007, 171.
48. Crick 1982, 58.
49. Sagan 1967; Margulis 1981.
50. Brockman 1996, 134.
51. Lovelock 1979.
52. Sampson 2012.
53. Brockman 1996, 87. Nonetheless, the Earth has evolved some of the self-regulating behaviors of a life form, which suggests that processes other than Darwinian selection may be at work.
54. Kauffman 1995.
55. Wolfram told *Forbes* magazine in 2000 that "I've come to believe that natural selection is not all that important." Malone 2000.
56. Lucretius 1951, 31.
57. Smolin 1997.
58. Actually, some believe that the existence of the universe can be proven mathematically. In an afterword to a book on the subject by physicist Lawrence Krauss, Richard Dawkins writes that "Even the last remaining trump card of the theologian, 'Why is there something rather than nothing?,' shrivels up before your eyes as you read these pages." Krauss 2012, 191.
59. Kaku 2005.
60. Smolin 2006b, 176.
61. Tolstoy 1977, 58.
62. Tolstoy 1977, 60–61.
63. Tolstoy 1977, 71.
64. Tolstoy 1977, 73.
65. In Wertheim 1997, 253. Of course, many scientists and philosophers from Pythagoras to Copernicus to Newton saw their work in predominately religious terms; but "in spite of the officially secular climate of modern science," according to Wertheim, "physicists have continued to retain a quasi-religious attitude to their work. They have continued to comport themselves as a scientific priesthood, and to present themselves to the public in that light."
66. Danchin 2002.
67. According to Marvin Minsky, one of the founders of artificial intelligence, "Scientists shouldn't have ethical responsibility for their inventions, they should be able to do

what they want. You shouldn't ask them to have the same values as other people." Quoted in Egan 2007. See Smolin 2006b, 303 for a discussion of ethics in science.

68. In one incident, a number of papers on string theory published in reputable journals were thought by some to be part of a hoax. They actually turned out to be genuine, just not comprehensible. One scientist wrote that "no one in the string group at Harvard can tell if these papers are real or fraudulent." Woit 2006, 218.

10. The Left-Hander's Guide to the Cosmos

1. Quoted in Moore 1967, 140.
2. Einstein, Podolsky, and Rosen 1935.
3. This version was proposed later by David Bohm. See Bohm 1951.
4. Quoted in Kumar 2008, 315.
5. Bell 1964.
6. Commonly attributed quotes are that quantum mechanics is "fundamentally incomprehensible" (Niels Bohr); "If you think you understand quantum mechanics, you don't understand quantum mechanics" (Richard Feynman); "You don't understand quantum mechanics, you just get used to it" (John von Neumann).
7. In Buchanan 2011.
8. Tversky and Shafir 1992.
9. See, for example, Kahneman 2011.
10. Henschen 1926.
11. Sperry 1975. See also Sperry and Gazzaniga 1967.
12. Inversions of the standard pattern may be associated with conditions such as schizophrenia, dyslexia, and autism, and "may confer special benefits, or lead to disadvantages, in the carrying out of different activities." McGilchrist 2009, 12.
13. McGilchrist 2009, 217.
14. Conditions such as autism are linked to deficiencies in right brain activity. According to Allan Snyder, the autistic mind concentrates on discrete detail, and so also "sees more of the parts than the whole." Snyder 2004. On the other hand, many of our most creative and successful scientists have showed some of the traits of autism. Fitzgerald 2004.
15. McGilchrist 2009, 193.
16. McGilchrist 2009, 85.
17. McGilchrist 2009, 82.
18. McGilchrist 2009, 127.
19. McGilchrist 2009, 82.
20. McGilchrist 2009, 57.
21. McGilchrist 2009, 98.
22. McGilchrist 2009, 174.
23. McGilchrist 2009, 209.
24. McGilchrist 2009, 209.
25. McGilchrist 2009, 235.
26. McGilchrist 2009, 6.
27. "Symmetry—in poetry, in music, in architecture, in prose and in thought—was perhaps the ultimately guiding aesthetic principle of the Enlightenment." McGilchrist 2009, 344.
28. Regard and Landis 1988.

29. B. Edwards 1999.
30. Orrell 2007, 320; Orrell 2008.
31. Midgley 2000, 14.
32. Knight 1921.
33. Crow 2006.
34. J.B. Taylor 2008, 142.
35. J.B. Taylor 2008, 69.
36. The theoretical physicist Anthony Zee wrote that physicists have learned to listen "to that half of the brain concerned with aesthetics. To read His mind, they search their own minds for that which constitutes symmetry and beauty. In the silence of the night, they listen for voices telling them about yet-undreamed-of symmetries." Both halves of the brain are concerned with aesthetics, but here it sounds like the left is doing the talking. Zee 1986, 99.
37. Einstein 1952.
38. Coulson and Williams 2005.
39. Nietzsche 1967, 97–98.
40. In Brown 1988, 108.
41. As Jacques Monod wrote, "There is and will remain a Platonic element in science which could not be taken away without ruining it. Among the infinite diversity of singular phenomena science can only look for invariants." Monod 1971, 100.
42. Einstein's thinking on the dynamics of spacetime was "helped by the movements he sensed within his arm as he squeezed a rubber ball." Peat 2008.
43. McGilchrist 2009, 54.
44. As pointed out in Chapter 9, while the multiverse theory takes an apparently pluralistic approach, its real aim is to explain the universe in terms of a unified theory by evoking Democritean randomness. It therefore favors unification of the model over anything to do with the actual world that we experience, and is another example of what Margaret Wertheim called the "quest for something utterly *disembodied*, something utterly *immaterial*" that characterises modern physics (Chapter 5).
45. Brockman 2004.
46. Koestler 1969. He introduced the concept of a holon in his book *The Ghost in the Machine*; see Koestler 1967.
47. See Waldrop 1992; Kauffman 1995.
48. Goldenfeld and Kadanoff 1999.
49. An extensive comparison of forecasting techniques used in areas such as economics found that "simple methods . . . do as well, or in many cases better, than statistically sophisticated ones." Makridakis and Hibon 2000. See also Granger and Jeon 2003. In systems biology, simple models of limited scope are much more useful for making predictions than complicated models that attempt to capture every detail of the system. See also the discussion of climate modeling in Chapter 8.
50. Orrell 2008, 199.
51. Borewein 2006, 23.
52. Orrell and Smith 2003.
53. Quoted in Burtt 2003.
54. Mandelbrot 1982.
55. Mandelbrot and Hudson 2004.
56. Penrose 2005, 756.
57. The distribution of price changes for major international stock market indices

follows a power-law distribution with a power of approximately three. See Lux 1996; Gopikrishnan et al. 1998; Gopikrishnan et al. 1999; Plerou et al. 1999.

58. Rucker 2012.
59. The Institute for Systems Biology was set up by Leroy Hood and others in 2000. Its guiding principles include integration of different technologies and disciplines; models that span the range of biological complexity from molecular to organism; open-source data and tools; and openness to outside collaboration. Hood 2008.
60. Guthrie 1962, 221.
61. Keynes 1936.
62. Soddy 1922, 6.
63. Mill 1907.
64. Haldane 2009.
65. May, Levin, and Sugihara 2008.
66. Vitali, Glattfelder, and Battiston 2011. Quote from Coghlan and MacKenzie 2011.
67. Ingram 2005, 211.
68. This optimistic bias may explain why, over the last thirty years, the average probability forecasters put on the economy lapsing into recession "has never risen above 50 percent—until the economy was already in a recession," according to the *New York Times*. Leonhardt 2011.
69. Some ideas are to introduce transaction taxes, increase reserve requirements, and strengthen financial firewalls. See Orrell 2010.
70. Gordon 2008, 208.
71. One project I am involved in which takes this approach aims to assess the effect of socio-demographics on travel demand over the next thirty years in the United States. See http://www.systemsforecasting.com.
72. Barfield 1988.
73. As McGilchrist notes, "Reading left to right involves moving the eyes towards the right, driven by the left hemisphere, and preferentially communicating what is seen to the left hemisphere." McGilchrist 2009, 276. The transition took place between the seventh and fifth centuries BC and included an intermediate stage known as *boustrophedon*, meaning "as the ox plows," in which the direction reversed at the end of each line.
74. See, for example, Markopoulou 2009.
75. Verlinde 2011.
76. In his book *The End of Science*, John Horgan punctured the pretensions of (among other things) chaos and complexity, which he treated as a single field called chaoplexity, and observed that one of its "most profound goals" was "the elucidation of a new law, or set of principles, or unified theory, or *something* that will make it possible to understand and predict the behavior of a wide variety of seemingly dissimilar complex systems." Horgan 1997, 277. I would argue that one of the main lessons of these fields, which is resisted even by some of its champions, is that there is no such unified theory or predictive model. This does not imply that the results are without interest or application. For example, we may not be able to predict a system, but we can learn how to make it more robust.
77. Laughlin 2005, 80.
78. Rukeyser 1968.
79. Sagan 1997, 131.
80. Feynman and Leighton 1988.

Bibliography

Abella, A. *Soldiers of reason: The Rand Corporation and the rise of the American empire.* Orlando: Harcourt, Inc., 2008.

Abraham, R. "The broken chain." 24 November 2005. http://www.ralph-abraham.org/articles/MS%23117.Kepler/kepler02.pdf (accessed 30 March 2012).

Bhikkhu, T. *Acintita Sutta Unconjecturable.* Vol. 3 in *Handful of Leaves: An Anthology From the Anguttara Nikaya.* Valley Center, CA: Sati Center for Buddhist Studies & Metta Forest Monastery, 2003.

Adams, D. *The original hitchhiker radio scripts.* Edited by G. Perkins. London: Pan Books, 1985.

Adler, R. "Ultimate guide to the multiverse." *New Scientist*, 28 November 2011.

Alberti, L.B. *The ten books of architecture: The 1755 Leoni edition.* Mineola, NY: Dover, 1986.

Alexander, S. *Space, time, and deity.* London: Macmillan, 1920.

Alighieri, D. *Il convito: The banquet of Dante Alighieri.* Translated by E.P. Sayer. London: G. Routledge and Sons, 1887.

Allen, W. *Getting even.* New York: Random House, 1971.

Alpher, R.A., H. Bethe, and G. Gamow. "The origin of chemical elements." *Physical Review* 73, no. 7 (1948): 803–804.

Amaldi, U., W. de Boer, and H. Furstenau. "Comparison of grand unified theories with electroweak and strong coupling constants measured at LEP." *Phys. Lett. B* 260 (1991): 447–455.

Amis, M. *The information.* New York: Vintage Books, 1995.

Anderson, C.D. "The positive electron." *Physical Review* 43, no. 6 (1933): 491–494.

Anderson, P.W. "More is different." *Science* 177, no. 4047 (1972): 393–396.

Anker, S., and D. Nelkin. *The molecular gaze: Art in the genetic age.* Cold Spring Harbor, NY: Cold Spring Harbor Laboratory Press, 2004.

Anonymous. "Enrico Fermi dead at 53; Architect of atomic bomb." *The New York Times*, 23 November 1954.

———. "1957: Sputnik satellite blasts into space." *BBC*, 4 October 1957. http://news.bbc.co.uk/onthisday/hi/dates/stories/october/4/newsid_2685000/2685115.stm (accessed 1 January 2012).

———. "Glimpses of superhistory." *CERN Courier*, 26 February 2001.

———. "Faces and places, Yuval Ne'eman." *CERN Courier*, 25 July 2006a: 6.

———. "Dismal science, dismal sentence." *Economist*, 9 September 2006b.

———. "248-dimension maths puzzle solved." *BBC News*, 19 March 2007. http://news.bbc.co.uk/1/hi/6466129.stm (accessed 25 November 2011).

———. "The other-worldly philosophers." *Economist*, 16 July 2009.

———. "Trader was not a hoaxer, says BBC." *BBC News*, 28 September 2011a. http://www.bbc.co.uk/news/business-15078419 (accessed 30 September 2011).

———. "Big chill 'on way.'" *The Sun*, 18 October 2011b.

———. "Neutrinos and multiverses: a new cosmology beckons." *New Scientist*, 28 November 2011c.

Appelbaum, B. "Inside the fed in '06: Coming crisis, and banter." *New York Times*, 13 January 2012.

Aristotle. *Aristotle's politics.* Translated by B. Jowett. New York: Modern Library, 1943.

———. *Metaphysics.* Translated by W.D. Ross. Sioux Falls, SD: NuVision Publications, 2009.

Arrow, K.J., and G. Debreu. "Existence of a competitive equilibrium for a competitive economy." *Econometrica* 22 (1954): 65–90.

Atkinson, N. "Q&A with Brian Cox, part 1: Recent hints of the Higgs." *Universe Today*, 16 August 2011. http://www.universetoday.com/88187/qa-with-brian-cox-part-1-recent-hints-of-the-higgs/ (accessed 1 September 2011).

Bachelier, L. "Théorie de la spéculation." *Annales Scientifiques de l'École Normale Supérieure* 3, no. 17 (1900): 21–86.

Bacon, F. "Of beauty." In *Bacon's essays: With annotations*, edited by R. Whately, 395–396. London: J.W. Parker & Son, 1856.

Bak, P. *How nature works: The science of self-organized criticality.* New York: Springer-Verlag, 1996.

Bakan, J. *The corporation: The pathological pursuit of profit and power.* New York: Free Press, 2004.

Barfield, O. *Saving the appearances: A study in idolatry.* 2nd Edition. Middletown: Wesleyan University Press, 1988.

Barr, S.M. *Modern physics and ancient faith.* Notre Dame, IN: University of Notre Dame Press, 2003.

Barres, B. "Does gender matter?" *Nature* 442 (2006): 133.

Barrow, J.D. *New theories of everything.* Oxford: Oxford University Press, 2007.

———. *Cosmic imagery: Key images in the history of science.* London: Bodley Head, 2008.

Bateson, M.C. "What is the most important invention in the past two thousand years? From: Mary Catherine Bateson." *Edge.org.* Edited by J. Brockman. 10 February 1999. http://edge.org/discourse/invention.html (accessed 8 January 2012).

Baxter, J. "Age of unreason." *Guardian*, 22 June 2004.

Beinhocker, E.D. *Origin of wealth: Evolution, complexity, and the radical remaking of economics.* Boston: Harvard Business School Press, 2006.

Bell, J.S. "On the Einstein-Podolsky-Rosen paradox." *Physics* 1 (1964): 195–200.

Bentham, J. *An introduction to the principles of morals and legislation.* Oxford: Clarendon Press, 1907.

Berger, J. *Ways of seeing.* London: British Broadcasting Corporation; Penguin Books, 1972.

Bernal, J.D. "Dr. Rosalind E. Franklin." *Nature* 182, no. 154 (1958).

Bernanke, B.S. "Commencement address at the Boston College School of Law." Newton, MA, 22 May 2009.

Beven, K. *Environmental modelling: An uncertain future.* London: Routledge, 2009.

Bizony, P. *Atom.* London: Icon, 2007.

Blake, W. "The Tyger." In *Songs of innocence and of experience: Shewing the two contrary states of the human soul.* London: W. Blake, 1794.

Bohm, D. *Quantum theory.* New York: Prentice-Hall, 1951.

———. *Causality and chance in modern physics.* Philadelphia: University of Pennsylvania, 1971.

———. *Wholeness and the implicate order.* London: Routledge, 1980.

Bohr, N. "On the constitution of atoms and molecules." *I. Phil. Mag.* 26 (1913): 1.

Borewein, J.M. "Aesthetics for the working mathematician." In *Mathematics and the aesthetic: New approaches to an ancient affinity*, edited by N. Sinclair, D. Pimm, and W. Higginson. New York: Springer, 2006.

Born, M. "Zur Quantenmechanik der Stoßvorgänge (Quantum mechanics of collision)." *Z. Phys.* 37 (1926): 863.

Bouchaud, J.-P. "Economics needs a scientific revolution." *Nature*, 2008: 1181.

Bowles, S. "Did warfare among ancestral hunter-gatherers affect the evolution of human social behaviors?" *Nature* 456 (2009): 326–327.

Braque, G. *Illustrated notebooks, 1917–1955.* Translated by S. Appelbaum. New York: Dover, 1971.

Brockman, J. *The third culture: Beyond the scientific revolution.* New York: Simon & Schuster, 1996.

———."A theory of roughness: A talk with Benoît Mandelbrot." *Edge*, 20 December 2004.http://www.edge.org/3rd_culture/mandelbrot04/mandelbrot04_index.html (accessed 16 January 2012).

Bronowski, J. *The common sense of science.* Harmondsworth: Penguin Books, 1960.

Brown, R.M. Starting from scratch: A different kind of writer's manual. Toronto: Bantam Books, 1988.

Browne, M.N., and J.K. Quinn. "The lamentable absence of power in mainstream economics." In *Future directions for heterodox economics*, edited by J.T. Harvey and R.F. Garnett. University of Michigan Press, 2008.

Brownlee, C. "Hubble's guide to the expanding universe." *PNAS.org.* 2011. http://www.pnas.org/site/misc/classics2.shtml (accessed 25 November 2011).

Brumfiel, G. "Beautiful theory collides with smashing particle data." *Nature* 471 (2011): 13–14.

Brummett, B. *Rhetoric of machine aesthetics.* Westport, CT: Praeger, 1999.

Buchanan, M. "Quantum minds: Why we think like quarks." *New Scientist*, 5 September 2011.

Buckley, P., and F.D. Peat. "Werner Heisenberg 1901–1976." In *Glimpsing reality: Ideas in physics and the link to biology*. Toronto: University of Toronto Press, 1996.

Buffet, W.E. "Chairman's 1988 letter to the shareholders of Berkshire Hathaway Inc." 28 February 1989.

Burkert, W. *Lore and science in ancient pythagoreanism.* Cambridge, MA: Harvard University Press, 1972.

Burnet, J. *Early Greek philosophy.* 3rd Edition. London: A. & C. Black Ltd, 1920.

Burtt, E.A. *The metaphysical foundations of modern science.* Mineola, NY: Dover, 2003.

Bushong, B., L.M. King, C.F. Camerer, and A. Rangel. "Pavlovian processes in consumer choice: The physical presence of a good increases willingness-to-pay." *American Economic Review* 100 (2010): 1–18.

Calder, N. *Magic universe: the Oxford guide to modern science.* Oxford: Oxford University Press, 2003.

Caldicott, H. "This beautiful planet." Minnesota Public Radio, 12 December 1981.

Campbell, J. *The masks of God: Occidental mythology.* Vol. 3. Harmondsworth: Penguin, 1991.

Campbell, K. *Reality on the rocks: Beyond our Ken*, London: Windfall Films, 1995.

Candelas, P., G.T. Horowitz, A. Strominger, and E. Witten. "Vacuum configurations for superstrings." *Nucl. Phys. B* 258, no. 1 (1985): 46–74.

Capra, F. *The Tao of physics: An exploration of the parallels between modern physics and eastern mysticism*. 2nd. Boulder, Colo.: Shambhala, 1983.

———. *The web of life: a new scientific understanding of living systems*. New York: Anchor Books, 1996.

Carey, N. *The epigenetics revolution: how modern biology is rewriting our understanding of genetics, disease, and inheritance*. London: Icon, 2011.

Caroe, L. "Britain faces a mini 'ice age'; This winter will see start of DECADES of big freezes." *The Daily Express*, 10 October 2011.

Cashford, J. *The moon: Myth and image*. New York: Four Walls Eight Windows, 2002.

Cathcart, B. *The fly in the cathedral: How a small group of Cambridge scientists won the race to split the atom*. New York: Farrar, Straus, and Giroux, 2004.

Chamberlain, O., E. Segre, C. Wiegand, and T. Ypsilantis. "Observation of antiprotons." *Physical Review* 100 (1955): 947–950.

Chandrasekhar, S. *Truth and beauty: Aesthetics and motivations in science*. Chicago: University of Chicago Press, 1987.

Clapworthy, G., M. Viceconti, P.V. Coveney, and P. Kohl. "The virtual physiological human: building a framework for computational biomedicine I. Editorial." *Philos. Transact A Math. Phys. Eng. Sci.* 366, no. 1878 (2008): 2975–2978.

Clement, D. "Interview with Eugene Fama." *The Region*, 1 December 2007.

Coghlan, A., and D. MacKenzie. "Revealed—the capitalist network that runs the world." *New Scientist*, 19 October 2011.

Copernicus, N. *On the revolutions*. Edited by J. Dobrzycki. Translated by E. Rosen. Baltimore: Johns Hopkins University Press, 1978.

Cornford, F.M. "Mysticism and science in the Pythagorean tradition." *The Classical Quarterly* 16 (1922): 137–150.

Coulomb, C. "Premier mémoire sur l'electricité et le magnétisme." *Histoire de l'Académie Royale des Sciences*, 1785: 569–577.

Coulson, S., and R.F. Williams. "Hemispheric asymmetries and joke comprehension." *Neuropsychologia* 43 (2005): 128–141.

Coumou, D., and S. Rahmstorf. "A decade of weather extremes." *Nature Climate Change*, 2012.

Crick, F. *Life itself: Its origin and nature*. New York: Simon and Schuster, 1982.

Crow, T.J. "March 27, 1827 and what happened later—the impact of psychiatry on evolutionary theory." *Progress in Neuro-Psychopharmacology and Biological Psychiatry* 30, no. 5 (2006): 785–796.

Crowe, K. "The Prediction Business." *The National*. CBC. Toronto. 24 March 2009.

Curtis, A. (director). *All watched over by machines of loving grace*. Produced by L. Kelsall. 2011.

Dalton, J. *Elements of English grammar*. London: W.J. and J. Richardson, 1801.

———. *A new system of chemical philosophy*. Manchester: S. Russell, 1808.

———. *Meteorological observations and essays*. 2nd Edition. Manchester: Printed by Harrison & Crosfield for Baldwin and Cradock, London, 1834.

Daly, H.E. *Beyond growth: The economics of sustainable development*. Boston, MA: Beacon Press, 1996.

Danby, G., et al. "Observation of high-energy neutrino reactions and the existence of two kinds of neutrinos." *Physical Review Letters* 9 (1962): 36.

Danchin, A. *The Delphic boat: What genomes tell us*. Translated by A. Quayle. Harvard: Harvard University Press, 2002.

Darwin, C. *The life and letters of Charles Darwin.* Edited by F. Darwin. London: Murray, 1887.

Davies, J.B., S. Sandstrom, A. Shorrocks, and E.N. Wolff. *The World Distribution of Household Wealth.* Helsinki: World Institute for Development Economics Research of the United Nations University, 2006.

Davies, P. *The Goldilocks enigma: Why is the universe just right for life?* London: Penguin, 2007.

Davis, N.P. *Lawrence and Oppenheimer.* New York: Simon and Schuster, 1968.

Dawkins, R. *The selfish gene.* New York: Oxford University Press, 1976.

———. "The ultraviolet garden." Royal Institution Christmas Lecture No 4. 1991.

———. *Unweaving the rainbow: Science, delusion, and the appetite for wonder.* Boston: Houghton Mifflin, 1998.

———. *The God delusion.* Boston: Houghton Mifflin Co., 2006.De Botton, A. *Status anxiety.* New York: Pantheon Books, 2004.

de Broglie, L. *Recherches sur la théorie des quanta.* PhD thesis. Paris: Masson, 1963.

Defense Advanced Research Projects Agency. *DARPA: 50 years of bridging the gap.* Tampa, FL: Faircount LLC, 2008.

Descartes, R. *Principles of philosophy (1644).* Vol. 1 in *The Philosophical Writings of Descartes*, translated by J. Cottingham, R. Stoothoff, and D. Murdoch. Cambridge: Cambridge University Press, 1985.

———. *Descartes: Selected philosophical writings.* Cambridge: Cambridge University Press, 1988.

Deutsch, D. *The fabric of reality.* London: Penguin, 1997.

DeWolfe, O., A. Giryavets, S. Kachru, and W. Taylor. "Type IIA moduli stabilization." *J. High Energy Physics* 503 (2005): 66.

Diacu, F. *Megadisasters: The science of predicting the next catastrophe.* Princeton: Princeton University Press, 2009.

Dirac, P.A.M. "The quantum theory of dispersion." *Proc. Roy. Soc. A* 114 (1927): 710.

———. "The quantum theory of the electron." *Proc. R. Soc. London A* 117 (1928): 610.

———. *The principles of "quantum" mechanics.* Oxford: Clarendon Press, 1930.

"Banquet speech." In *Les Prix Nobel en 1933*, edited by C.G. Santesson. Stockholm: Nobel Foundation, 1934.

———. "The evolution of the physicist's picture of nature." *Scientific American*, May 1963.

Distler, J., and S. Garibaldi. "There is no 'theory of everything' inside E8." *J. Math. Phys.* 298 (2010): 419–436.

Douglas, M.R. "The statistics of string/M theory vacua." *J. High Energy Phys.* 5 (2003): 46.

Doward, J., and B. Wilkinson. "Are beautiful people 'selfish by nature'?" *The Observer*, 14 August 2011.

Dryden, J. "A song for St. Cecilia's Day, 1687." In *The poetical works of John Dryden*, 78. Edinburgh: J. Nichol, 1855.

Du Fay, C. "A discourse concerning electricity." *Philosophical Transactions of the Royal Society* 38 (1734): 258–266.

Duff, M., and C. Sutton. "The membrane at the end of the universe." *New Scientist*, 30 June 1988: 67.

Dürer, A. *The painter's manual: A manual of measurement of lines, areas, and solids by means of compass and ruler assembled by Albrecht Dürer for the use of all lovers of art with appropriate illustrations arranged to be printed in the year MDXXV.* Translated by W.L. Strauss. New York: Abaris Books, 1977.

Durrell, L. *Balthazar.* New York: Dutton, 1958.

Dyson, F. "Obituary of Hermann Weyl." *Nature* 177 (1956): 457–458.

Eco, U. *The aesthetics of Thomas Aquinas.* Cambridge, MA: Harvard University Press, 1988.

Edgeworth, F.Y. *Mathematical psychics: An essay on the application of mathematics to the moral sciences.* London: C.K. Paul, 1881.

Editorial. "Einstein=Man of Conscience2." *Scientific American*, September 2004: 10.

Edwards, B. *Drawing on the right side of the brain.* London: HarperCollins, 1999.

Edwards, P.N. "A brief history of atmospheric general circulation modeling." In *General Circulation Model Development*, edited by D.A. Randall, 67–90. San Diego, CA: Academic Press, 2000.

Egan, D. "The plan for eternal life." *New Scientist*, 13 October 2007.

Einstein, A. "Über einen die Erzeugung und Verwandlung des Lichtes betreffenden heuristischen Gesichtspunkt (On a heuristic viewpoint concerning the production and transformation of light)." *Annalen Der Physik* 17, no. 6 (1905): 132–148.

———. "Die Feldgleichungen der Gravitation." In *Sitzungsberichte*, 778–786. Berlin: Königlich Preußische Akademie der Wissenschaften, 1915.

———. "Die Grundlage der allgemeinen Relativitätstheorie (The foundation of the general theory of relativity)." *Annalen Der Physik* 49, no. 7 (1916): 769–822.

———. "My first impressions of the U.S.A. " *Nieuwe Rotterdamsche Courant*, 4 July 1921.

———. "Does the inertia of a body depend upon its energy-content?" In *The principle of relativity*, translated by G.B. Jeffery and W. Perrett. London: Methuen and Company Ltd., 1923a.

———. "On the electrodynamics of moving bodies." In *The principle of relativity*, edited by G.B. Jeffery and W. Perrett. London: Methuen and Company Ltd., 1923b.

———. "Das Komptonsche Experiment (The Compton experiment)." *Berliner Tageblatt*, 20 April 1924.

———. "Religion and science." *New York Times Magazine*, 9 November 1930: 1–4.

———. *On the method of theoretical physics: The Herbert Spencer lecture.* Oxford: Clarendon Press, 1933.

———. "A letter to Jacques Hadamard." In *The creative process*, edited by B. Ghiselin, 32. Berkeley, CA: University of California Press, 1952.

———. "Letter to President Franklin P. Roosevelt, (2 Aug 1939, delivered 11 Oct 1939)." In *Einstein on peace*, edited by O. Nathan and H. Norden, 294–295. New York: Simon and Schuster, 1960.

———. *The collected papers of Albert Einstein.* Vol. 8. Princeton, NJ: Princeton University Press, 1987.

———. "Principles of research." In *Einstein's essays in science*. Mineola, NY: Dover, 2009.

Einstein, A., B. Podolsky, and N. Rosen. "Can quantum-mechanical description of physical reality be considered complete?" *Phys. Rev.* 47 (1935): 777.

Else, J.H. (director). *The day after Trinity.* Produced by KTEH public television. 1980.

Euclid. *The thirteen books of Euclid's elements.* Translated by T.L. Heath. Cambridge: Cambridge University Press, 1908.

European Society of Human Genetics. "Direct-to-consumer genetic tests neither accurate in their predictions nor beneficial to individuals, European geneticists say." 31 May 2011. https://www.eshg.org/13.0.html (accessed 30 March 2012).

Everett, H. "Relative state formulation of quantum mechanics." *Rev. Mod. Phys.* 29 (1957): 454–462.

Fama, E.F. *Random walks in stock-market prices.* Chicago: Graduate School of Business, University of Chicago, 1965.

Faraday, M. *Experimental researches in electricity.* Vol. 3. London: Taylor, 1855.

Farmelo, G. *The strangest man: The hidden life of Paul Dirac, quantum genius.* London: Faber, 2009.

Farrell, T.F. "Trinity test, July 16, 1945. Eyewitness Brigadier General Thomas F. Farrell." *NuclearFiles.org.* 1945. http://www.nuclearfiles.org/menu/key-issues/nuclear-weapons/history/pre-cold-war/manhattan-project/trinity/eyewitness-thomas-farrell_1945-07-16.htm (accessed 31 December 2011).

Ferguson, K. *Pythagoras: His lives and the legacy of a rational universe.* London: Icon, 2010.

Fermi, L. *Atoms in the family: My life with Enrico Fermi.* Chicago: University of Chicago Press, 1954.

Feynman, R. "Space-time approach to non-relativistic quantum mechanics." *Rev. of Mod. Phys.* 20 (1948): 367.

———. *The Feynman lectures on physics.* Vol. 1. Reading, MA: Addison Wesley, 1964.

———. *The pleasure of finding things out: The best short works of Richard P. Feynman.* New York: Basic Books, 2000.

Feynman, R., and R. Leighton. *What do YOU care what other people think?: Further adventures of a curious character.* New York: Norton, 1988.

Fildes, R., and N. Kourentzes. "Validation and forecasting accuracy in models of climate change." *International Journal of Forecasting* 27 (2011): 968–995.

Finkbeiner, A.K. *The Jasons: The secret history of science's postwar elite* . New York: Viking, 2006.

Fitzgerald, M. *Autism and creativity: Is there a link between autism in men and exceptional ability?* Hove, NY: Brunner-Routledge, 2004.

Fludd, R. *Philosophia sacra et vere christiana seu meteorologia cosmica.* Frankfurt: Francofurti prostat in officina Bryana, 1626.

Forrester, J.W. "Economic theory for the new millennium." *International System Dynamics Conference.* New York, 21 July 2003.

Fraser, A. "L'1%, c'est moi." *Texte Zur Kunst,* September 2011: 114.

Friedman, M. *Essays in positive economics.* Chicago: University of Chicago Press, 1953.

Frisch, O.R. *What little I remember.* Cambridge: Cambridge University Press, 1979.

Fry, R.E. *Transformations.* London: Chatto & Windus, 1926.

Gaither, C.C., and A.E. Cavazos-Gaither. *Gaither's dictionary of scientific quotations.* New York: Springer, 2012.

Galilei, G. *Dialogues concerning two new sciences.* Translated by H. Crew and A. de Salvio. New York: Dover, 1914.

———. *Dialogue concerning the two chief world systems.* Translated by S. Drake. Berkeley, CA: University of California Press, 1953.

Galilei, V. *Dialogue on ancient and modern music.* Edited by C.V. Palisca. New Haven, CT: Yale University Press, 2003.

Gamow, G. *My world line: An informal autobiography.* New York: Viking Press, 1970.

Gardner, M. "The fantastic combinations of John Conway's new solitaire game 'life.'" *Scientific American,* October 1970: 120–123.

Gartner, J. "Virtual Vermin Saves Lab Rats." *Wired,* 5 May 2005.

Gell-Mann, M. "A Schematic Model of Baryons and Mesons." *Phys. Lett.* 8 (1964): 214.

———. *Murray Gell-Mann on beauty and truth in physics.* March 2007. http://www.ted.com/talks/murray_gell_mann_on_beauty_and_truth_in_physics.html (accessed 9 December 2011).

Georgi, H., and S.L. Glashow. "Unity of all elementary-particle forces." *Phys. Rev. Lett.* 32 (1974): 438–441.

Ghosh, P. "LHC results put supersymmetry theory 'on the spot.'" *BBC News,* 27 August 2011. http://www.bbc.co.uk/news/science-environment-14680570 (accessed 1 September 2011).

Gilby, T. (translator). *St. Thomas Aquinas: Philosophical texts.* London: Oxford University Press, 1951.

Glashow, S. "The hunting of the quark." *The New York Times Magazine,* 18 July 1976.

———. "Partial-symmetries of weak interactions." *Nuclear Physics* 22 (1961): 579–588.

Gleick, J. *Chaos.* London: Viking, 1987.

Gleiser, M. *A tear at the edge of creation: A radical new vision for life in an imperfect universe.* New York: Free Press, 2010.

Goetzmann, W., L. Renneboog, and C. Spaenjers. *Art and money.* Working Paper, Yale School of Management , 2009.

Goldenfeld, N., and L.P. Kadanoff. "Simple lessons from complexity." *Science* 284 (1999): 87–89.

Gopikrishnan, P., et al. "Inverse cubic law for the distribution of stock price variations." *European Physical Journal B* 3 (1998): 139–140.

———. "Scaling of the distribution of fluctuations of financial market indices." *Physical Review E* 60 (1999): 5305–5316.

Gordon, A. *Future savvy: Identifying trends to make better decisions, manage uncertainty, and profit from change.* New York: American Management Association, 2008.

Granger, C.W.J., and Y. Jeon. "Interactions between large macro models and time series analysis." *International Journal of Finance & Economics* 8 (2003): 1–10.

Gray, L. "Met Office drops seasonal forecast." *Telegraph,* 5 March 2010.

Greene, B. "Welcome to the multiverse." *Newsweek,* 21 May 2012.

Green, M.B., and J.H. Schwarz. "Anomaly cancellations in supersymmetric D = 10 gauge theory and superstring theory." *Physics Letters B* 149, no. 1–3 (1984): 117–122.

Green, T. "Lourdes Benería." *Adbusters,* 15 July 2009.

Greenspan, A. "Testimony of Dr Alan Greenspan." *House Committee of Government Oversight and Reform.* Washington, DC, 23 October 2008.

Gribbin, J. *The scientists: a history of science told through the lives of its greatest inventors.* New York: Random House, 2002.

Gross, D.J. "Einstein and the search for unification." *Current Science* 89, no. 12 (December 2005).

———. "Gauge theory—past, present, and future?" *Chinese Journal of Physics* 30 (1992): 955.

Gross, D.J., and F. Wilczek. "Ultraviolet behaviour of non-abelian gauge theory." *Phys. Rev. Lett.* 30 (1973): 1343.

Guerrilla Girls. *Confessions of the Guerrilla Girls.* New York: Harper & Collins, 1995.

Guth, A. "Inflationary universe: A possible solution to the horizon and flatness problems." *Physical Review D* 23, no. 2 (1981): 347–356.

Guthrie, W.K.C. *A History of Greek Philosophy, The earlier Presocratics and the Pythagoreans.* Vol. 1. 6 vols. Cambridge: Cambridge University Press, 1962.

———. *A History of Greek Philosophy, The later Plato and the academy.* Vol. 5. 6 vols. Cambridge: Cambridge University Press, 1978.

Hafele, J., and R. Keating. "Around the world atomic clocks: Predicted relativistic time gains." *Science* 177 , no. 4044 (1972): 166–168.

Haldane, A.G. "Rethinking the financial network." Speech delivered at the Financial Student Association, Amsterdam, April 2009.

Halpern, P. *Collider: The search for the world's smallest particles.* Hoboken, NJ: John Wiley & Sons, 2009.

Hannam, J. *God's philosophers: How the medieval world laid the foundations of modern science.* London: Icon Books, 2009.

Hardy, G.H. *A mathematician's apology*. Cambridge: Cambridge University Press, 1940.

Häring, N. *Markt und Macht: Was Sie schon immer über die Wirtschaft wissen wollten, aber bisher nicht erfahren sollten.* Stuttgart: Schäffer-Poeschel, 2010.

Harris. "Most Americans think devastating natural disasters are increasing." Harris Interactive, 7 July 2011.

Harwood Group. *Yearning for balance: Views of Americans on consumption, materialism, and environment.* Takoma Park, MD: Merck Family Fund, 1995.

Hawking, S. *A brief history of time: From the big bang to black holes.* New York: Bantam Books, 1988.

Hawking, S., and L. Mlodinow. *The grand design.* New York: Bantam Books, 2010.

Heims, S. *John von Neumann and Norbert Wiener: From mathematics to the technologies of life and death.* Cambridge, MA: MIT Press, 1980.

Heisenberg, W. "Die Plancksche Entdeckung und die philosophischen Grundfragen der Atomlehre." *Die Naturwissenschaften* 45, no. 10 (1958): 227–234.

———. *Physics and beyond: Encounters and conversations.* New York: Harper & Row, 1971.

Heisenberg, W., and E. Heisenberg. *Meine liebe Li! Der Briefwechsel 1937–1946.* Edited by A.M. Hirsch-Heisenberg. St. Pölten: Residenz, 2011.

Henderson, L.D. *The fourth dimension and non-euclidean geometry in modern art.* Princeton, NJ: Princeton University Press, 1983.

Henschen, S.E. "On the function of the right hemisphere of the brain in relation to the left in speech, music and calculation." *Brain* 49 (1926): 110–126.

Heraclitus. *Fragments: The collected wisdom of Heraclitus.* Translated by B. Haxton. New York: Viking, 2001.

Higgs, P.W. "Broken Symmetries and Masses of Gauge Bosons." *Phys. Rev. Lett.* 13 (1964): 508.

Hillman, J. *Re-Visioning Psychology.* New York: Harper & Row, 1975.

Hinton, C.H. *Scientific romances: First and second series.* New York: Arno Press, 1976.

Hoffmann, B. *Albert Einstein: Creator and Rebel.* New York: Viking, 1972.

Holton, G.J., and S.G. Brush. *Physics, the human adventure: From Copernicus to Einstein and beyond.* New Brunswick, NJ: Rutgers University Press, 2001.

Holton, G.J., and Y. Elkana. *Albert Einstein, historical and cultural perspectives.* Princeton, NJ: Princeton University Press, 1982.

Hood, L. "A personal journey of discovery: Developing technology and changing biology." *Annual Review of Analytical Chemistry* 1 (2008): 1–43.

Horgan, J. *The end of science: Facing the limits of knowledge in the twilight of the scientific age.* New York: Broadway Books, 1997.

Horkheimer, M., and T.W. Adorno. *Dialectic of enlightenment: Philosophical fragments.* Translated by E. Jephcott. Stanford, CA: Stanford University Press, 2002.

Hoyle, F. "The universe: Past and present reflections." *Annual Review of Astronomy and Astrophysics* 20 (1982): 1–35.

Hutcheson, F. "An initial theory of taste." In *Aesthetics: A critical anthology*, edited by G. Dickie and R. J. Sclafani, 569–591. New York: St. Martin's Press, 1977.

Huxley, A. "Wordsworth in the tropics." In *Do what you will: Essays.* London: Chatto & Windus, 1929.

Iamblichus. *The Life of Pythagoras.* Translated by T. Taylor. Kila, MT: Kessinger, 1918.

Ingram, J. *Theatre of the Mind.* Toronto: HarperCollins, 2005.

IPCC. *Climate change 2001: The scientific basis.* Edited by J.T. Houghton, Y. Ding, D.J. Griggs, M. Noguer, P.J. van der Linden and D. Xiaosu. Cambridge: Cambridge University Press, 2001.

———. *Climate change 2007: The physical science basis. Contribution of Working Group I to the Fourth Assessment Report of the Intergovernmental Panel on Climate Change.* Edited by S. Solomon et al. Cambridge: Cambridge University Press, 2007.

Isaacson, W. *Einstein: His life and universe.* New York: Simon & Schuster, 2007.

Ivie, R., and C.L. Tesfaye. "Women in physics: A tale of limits." *Physics Today* 65, no. 2 (2012): 47.

James, O. *The selfish capitalist: Origins of affluenza.* London: Vermilion, 2008.

Janssens, A.C.J.W., M. Gwinn, L.A. Bradley, B.A. Oostra, C.M. van Duijn, and M.J. Khoury. "A critical appraisal of the scientific basis of commercial genomic profiles used to assess health risks and personalize health interventions." *American Journal of Human Genetics* 82, no. 3 (March 2008): 593–599.

Jayakumar, R. *Particle accelerators, colliders, and the story of high energy physics: Charming the cosmic snake* . New York: Springer Verlag, 2012.

Jevons, W.S. *The theory of political economy.* 5th Edition. New York: Kelley and Millman, 1957.

Johnson, G. *The ten most beautiful experiments.* New York: Alfred A. Knopf, 2008.

Joule, J.P. "On the mechanical equivalent of heat." *Brit. Assoc. Rep., Trans. Chemical Sect.*, 1845: 31.

Joyce, J. *A portrait of the artist as a young man.* New York: Penguin, 2003.

Kahneman, D. *Thinking, fast and slow.* New York: Farrar, Straus and Giroux, 2011.

Kahneman, D., and A. Tversky. "Prospect theory: An analysis of decision under risk." *Econometrica* 47 (1979): 263–291.

Kaku, M. "Unifying the Universe." *New Scientist*, 16 April 2005.

Kaptchuk, T.J., C.E. Kerr, and A. Zanger. "The art of medicine: Placebo controls, exorcisms, and the devil." *Lancet* 374, no. 9697 (2009): 1234–1235.

Kauffman, S. *At home in the universe: The search for the laws of self-organization and complexity.* Oxford: Oxford University Press, 1995.

Keats, J. "Ode on a Grecian Urn." In *Selected poems*, edited by J. Barnard, 191. New York: Penguin Books, 2007.

Keenlyside, N.S. "Commentary on 'Validation and forecasting accuracy in models of climate change.'" *International Journal of Forecasting* 27 (2011): 1000–1003.

Keepin, B., and B. Wynne. "Technical analysis of IIASA energy scenarios." *Nature* 312 (1984): 691–695.

Keller, E.F. *Reflections on gender and science.* New Haven, CT: Yale University Press, 1985.

Kepler, J. *Astronomiae pars optica (The optical part of astronomy).* Frankfurt: Apud Claudium Marnium & Hæredes Ioannis Aubrii, 1604.

Kerr, R.A. "Three degrees of consensus." *Science* 305 (2004): 932–934.

———. "How hot will the greenhouse world be?" *Science* 309 (2005): 100.

Kevles, D.J. "Good-bye to the SSC: On the life and death of the Superconducting Super Collider." *Engineering and science* 58, no. 2 (1995a): 16.

———. *The physicists: The history of a scientific community in modern America.* Cambridge, MA: Harvard University Press, 1995b.

Keynes, J.M. *The general theory of employment, interest and money.* New York: Harcourt, Brace, 1936.

Kirkby, J. et al. "Role of sulphuric acid, ammonia and galactic cosmic rays in atmospheric aerosol nucleation." *Nature* 476 (2011): 429–433.

Klein, N. *The shock doctrine: The rise of disaster capitalism* . Toronto: Alfred A. Knopf Canada, 2007.

Knight, F.H. *Risk, uncertainty, and profit.* Boston: Houghton Mifflin, 1921.

Koestler, A. *The sleepwalkers: A history of man's changing vision of the Universe.* London: Penguin, 1989.

———. *The ghost in the machine.* London: Hutchinson, 1967.

———. "Some general properties of self-regulating open hierarchic order (SOHO)." In *Beyond reductionism: New perspectives in the life sciences*, edited by A. Koestler and J.R. Smythies. London: Hutchinson, 1969.

Kollewe, J. "Pharmaceuticals struggle to find next blockbuster drugs as R&D costs soar." *Guardian*, 21 November 2011.

Koren, L. *Wabi-Sabi for artists, designers, poets and philosophers.* Berkeley, CA: Stone Bridge Press, 1994.

Korsmeyer, C. "Feminist aesthetics." *The Stanford encyclopedia of philosophy (Fall 2008 edition).* Edited by E.N. Zalta. 21 September 2008. http://plato.stanford.edu/archives/fall2008/entries/feminism-aesthetics/ (accessed 1 September 2011).

Krause, C. "The virtual human project: An idea whose time has come?" *Oak Ridge National Laboratory Review* 33, no. 1 (2000): 8–11.

Kroehling, R. (director). *A. Einstein: How I see the world.* Produced by K. D'Amico and L. Nathanson. 1991.

Krugman, P. "How did economists get it so wrong?" *New York Times*, 2 September 2009.

Kuhn, T.S. *The structure of scientific revolutions.* Chicago: University of Chicago Press, 1962.

Kumar, M. *Quantum: Einstein, Bohr and the great debate about the nature of reality.* London: Icon, 2008.

Laertius, D. *The lives and opinions of eminent philosophers.* Translated by C.D. Yonge. London: Henry G. Bohn, 1853.

Lasn, K., and Adbusters. *Occupy Econ 101: A manifesto for students all over the world.* New York: Seven Stories Press, 2012.

Laughlin, R.B. *A different universe: Reinventing physics from the bottom down.* New York: Basic Books, 2005.

Laurence, W. "Drama of the atomic bomb found climax in July 16 test." *New York Times*, 26 September 1945.

Lavoisier, A.L. *Elements of chemistry.* New York: Dover, 1965.

Lawrence, D.H. *Reflections on the death of a porcupine and other essays.* Philadelphia: Centaur Press, 1925.

Le Corbusier. *The city of to-morrow and its planning*. Translated by F. Etchells. Cambridge, MA: MIT Press, 1971.

Lederman, L., and D. Teresi. *The God particle: If the universe is the answer, what is the question?* New York: Houghton Mifflin, 1993.

Leibniz, G.W. *Discourse on metaphysics*.Chicago: The Open Court, 1902.

Lemonick, M. "Edward Witten." *TIME*, 26 April 2004.

Leonhardt, D. "Rising fears of recession." *New York Times*, 7 September 2011.

Levins, R. "The strategy of model building in population biology." *American Scientist* 54, no. 4 (1966): 421–431.

Levy, S. "The man who cracked the code to everything . . ." *Wired*, 1 June 2002: 132–148.

Lewis, D.K. *On the plurality of worlds*. Oxford: Blackwell, 1986.

Lewis, J.M. "Roots of ensemble forecasting." *Monthly Weather Review* 133, no. 7 (2005): 1865–1885.

Linde, A.D. "Eternally existing self-reproducing chaotic inflationary universe." *Physics Letters B* 175, no. 4 (1986): 395–400.

———. "2012 : What is your favorite deep, elegant, or beautiful explanation?" *Edge.org*. 16 January 2012. http://www.edge.org/responses/what-is-your-favorite-deep-elegant-or-beautiful-explanation (accessed 16 January 2012).

Lindley, D. *The end of physics: The myth of a unified theory*. New York: Basic Books, 1993.

———. *Uncertainty: Einstein, Heisenberg, Bohr, and the struggle for the soul of science*. New York: Anchor, 2007.

Lisi, A.G., and J.O. Weatherall. "A geometric theory of everything." *Scientific American*, 29 November 2010: 54–61.

Lo, A., and C. MacKinlay. "Stock market prices do not follow random walks: Evidence from a simple specification test." *Review of Financial Studies* 1 (1988): 41–66.

Lorenz, E. "Deterministic nonperiodic flow." *J. Atmos. Sci.* 20 (1963): 130–141.

———. "The butterfly effect." In *The chaos avant-garde: Memories of the early days of chaos theory*, edited by R. Abraham and Y. Ueda, 91–94. River Edge, NJ: World Scientific, 2000.

Lovejoy, A.O. *The great chain of being: A study of the history of an idea*. Cambridge, MA: Harvard University Press, 1936.

Lovelock, J.E. *Gaia: A new look at life on Earth*. Oxford: Oxford University Press, 1979.

Lucas, P., and A. Sheeran. "Asperger's syndrome and the eccentricity and genius of Jeremy Bentham." *Journal of Bentham Studies* 8 (2006): 1–37.

Lucretius. *On the nature of things*. Translated by R. Latham. Baltimore, MD: Penguin, 1951.

Lux, T. "The stable paretian hypothesis and the frequency of large returns: An examination of major German stocks." *Applied Financial Economics* 6 (1996): 463–475.

Magueijo, J. *Faster than the speed of light: The story of a scientific speculation*. Cambridge, MA: Perseus, 2003.

Mainzer, K. *Symmetries of nature: A handbook for philosophy of nature and science*. New York: Walter de Gruyter, 1996.

Makridakis, S., and M. Hibon. "The M3-Competition: Results, conclusions and implications." *International Journal of Forecasting* 16 (2000): 451–476.

Makridakis, S., and N.N. Taleb. "Decision making and planning under low levels of predictability." *International Journal of Forecasting* 25 (2009): 716–733.

Makridakis, S., R.M. Hogarth, and A. Gaba. "Forecasting and uncertainty in the economic and business world." *International Journal of Forecasting* 25 (2009): 794–812.

Malone, M.S. "God, Stephen Wolfram, and Everything Else." *Forbes ASAP*, 27 November 2000.

Malthus, T.R. *An essay on the principle of population.* Edited by A. Flew. Harmondsworth: Penguin, 1970.

Mandelbrot, B. *The Fractal Geometry of Nature.* New York: W. H. Freeman & Co., 1982.

Mandelbrot, B., and R.L. Hudson. *The misbehavior of markets: A fractal view of financial turbulence.* New York: Basic Books, 2004.

Mann, T. *The magic mountain.* Translated by J.E. Woods. New York: Vintage International, 1996.

Margulis, L. *Symbiosis in cell evolution: life and its environment on the early Earth.* San Francisco: W.H. Freeman, 1981.

Markopoulou, F. "New directions in background independent quantum gravity." In *Approaches to quantum gravity: Toward a new understanding of space, time, and matter*, edited by D. Oriti, 129. Cambridge: Cambridge University Press, 2009.

Matthews, R. "Don't blame the butterfly." *New Scientist*, 4 August 2001: 24.

———. "Nothing gained in search for 'theory of everything.'" *Financial Times*, 2 June 2006 .

Mattick, J.S. "The functional genomics of noncoding RNA." *Science* 309 (2005): 1525–1526.

Maxwell, J.C. "A dynamical theory of the electromagnetic field." *Philosophical Transactions of the Royal Society of London* 155 (1865): 459–512.

May, R.M., S.A. Levin, and G. Sugihara. "Ecology for bankers." *Nature* 451 (2008): 891–893.

McCauley, J.L. *Dynamics of markets: Econophysics and finance.* Cambridge: Cambridge University Press, 2004.

McCloskey, D. *Crossing: A memoir.* Chicago: University of Chicago Press, 1999.

———. "Crossing economics." *International Journal of Transgenderism* 4, no. 3 (2000).

McCloskey, D., and S.T. Ziliak. *Measurement and meaning in economics: The essential Deirdre McCloskey.* Cheltenham: Edward Elgar, 2001.

McClure, S.M., D.I. Laibson, G. Loewenstein, and J.D. Cohen. "Separate neural systems value immediate and delayed monetary rewards." *Science* 306, no. 5695 (2004): 503–507.

McGilchrist, I. *The master and his emissary.* New Haven, CT: Yale University Press, 2009.

McIlroy, A. "Hawking Centre opens to foster future Einsteins in Waterloo." *Globe and Mail*, 16 September 2011.

McMaster, J. *Viewpoints on string theory: Sheldon Glashow. NOVA online*, 2003a.

———. *Viewpoints on string theory: Edward Witten. NOVA online*, 2003b.

McSharry, P.E. "Validation and forecasting accuracy in models of climate change: Comments." *International Journal of Forecasting* 27 (2011): 996–999.

Mehra, J., and H. Rechenberg. *The historical development of quantum theory.* New York: Springer, 2000.

Mendeleev, D. *The Principles of Chemistry.* Longmans: London, 1891.

Mermin, N. David. "What's Wrong with this Pillow?" *Physics Today*, April 1989: 9.

Mesmer, F.A. *Mesmerism: Being the first translation of Mesmer's historic Mémoire sur la découverte du magnétisme animal, to appear in English.* Translated by G. Frankau. London: Macdonald, 1948.

Metzner, P. *Crescendo of the virtuoso: Spectacle, skill, and self-promotion in Paris during the Age of Revolution.* Berkeley, CA: University of California Press, 1998.

Midgley, M. *Evolution as a religion: Strange hopes and stranger fears.* London: Methuen, 1985.

———. *Gaia: The next big idea.* London: Demos, 2000.

Mill, J.S. "On social freedom." *Oxford and Cambridge Review* 1 (January 1907).

Millay, E.S.V. *The ballad of the harp-weaver.* New York: Frank Shay, 1922.

Miller, A.I. *Einstein, Picasso: Space, time and the beauty that causes havoc.* New York: Basic Books, 2002.\

———. "Science: A thing of beauty." *New Scientist*, 4 February 2006: 50.

Minkowski, H. "Space and time." In *The principle of relativity: A collection of original memoirs on the special and general theory of relativity*, edited by H.A. Lorentz, 75–91. New York: Dover, 1952.

Minsky, H.P. *Financial instability revisited: The economics of disaster.* Washington, DC: Board of Governors of the Federal Reserve System, 1972.

Moerman, D.E. *Meaning, medicine, and the 'placebo effect'.* Cambridge: Cambridge University Press, 2002.

Monbiot, G. "That sleighbell winter? It's all part of climate change denial." *Guardian*, 2 January 2012.

Monod, J. *Chance and necessity: An essay on the natural philosophy of modern biology.* New York: Knopf, 1971.

Moore, R. "Frank Gehry: 'There's a backlash against me.'" *The Observer*, 19 February 2012.

Moore, R.E. *Niels Bohr, the man and the scientist.* London: Hodder & Stoughton, 1967.

Moore, W.J. *Schrödinger, life and thought.* Cambridge: Cambridge University Press, 1989.

Morgan, T.H., A.H. Sturtevant, H.J. Muller, and C.B. Bridges. *The mechanism of Mendelian heredity.* New York: Henry Holt, 1915.

Motl, L. "Beauty of string theory." *The Reference Frame,* 25 October 2004. http://motls.blogspot.com/2004/10/beauty-of-string-theory.html (accessed 13 December 2011).

Mullin, T. "Turbulent times for fluids." *New Scientist*, 11 November 1989.

Nelson, J.A. "The masculine mindset of economic analysis." *The Chronicle of Higher Education* 42 (1996): B3.

Newlands, J.A.R. "On the law of octaves." *Chemical News* 12 (1865): 83.

Newton, I. *Opticks: Or, a treatise of the reflections, refractions, inflections and colours of light.* London: William Innys, 1730.

———. "Fragments from a treatise on revelation." In Frank E. Manuel, *The religion of Isaac Newton*, 120. Oxford: Clarendon Press, 1974.

Newton-Smith, W.H. *A companion to the philosophy of science.* Oxford: Blackwell, 2000.

Nicolson, M.H. *Mountain gloom and mountain glory: The development of the aesthetics of the infinite.* Ithaca, NY: Cornell University Press, 1959.

Nielsen. "Global concern for climate change dips amid other environmental and economic concerns." 29 August 2011. http://blog.nielsen.com/nielsenwire/consumer/global-concern-for-climate-change-dips-amid-other-environmental-and-economic-concerns/

Nietzsche, F. *Beyond good and evil: Prelude to a philosophy of the future.* Translated by W. Kaufmann. New York: Random House, 1966.

———. *The birth of tragedy, and the case of Wagner.* Translated by W. Kaufmann. New York: Vintage Books, 1967.

Noble, H.B. "Milton Friedman, 94, free-market theorist, dies." *New York Times*, 17 November 2006.

Nochlin, L. "Why have there been no great women artists?" In *Women, Art and Power*, 145–178. New York: Harper & Row, 1988.

Noether, E. "Invariante variationsprobleme (Invariant variation problems)." *Nachr. v. d. Ges. d. Wiss. zu Göttingen*, 1918: 235–257.

Nuwer, R. "How much will the Earth warm up?" *Green, New York Times blog*, 24 November 2011. http://green.blogs.nytimes.com/2011/11/24/how-much-will-the-earth-warm-up/ (accessed 12 December 2011).

Oppenheimer, J.R. *Science and the common understanding.* New York: Simon and Schuster, 1954.

———. "J. Robert Oppenheimer on the Trinity test." *Atomic Archive.* 1965. http://www.atomicarchive.com/Movies/Movie8.shtml (accessed 29 December 2011).

Orrell, D. *Modelling nonlinear dynamical systems: Chaos, error, and uncertainty.* D.Phil. Thesis. Oxford: University of Oxford, 2001.

———. "Role of the metric in forecast error growth: How chaotic is the weather?" *Tellus* 54A (2002): 350–362.

———. *Apollo's arrow: The science of prediction and the future of everything.* Toronto: HarperCollins, 2007.

———. *The other side of the coin: The emerging vision of economics and our place in the world.* Toronto: Key Porter, 2008.

———. *Economyths: How the science of complex systems is transforming economic thought.* London: Icon, 2010.

Orrell, D., L. Smith, J. Barkmeijer, and T. Palmer. "Model error in weather forecasting." *Nonlinear Proc. Geoph.* 9 (2001): 357–371.

Orrell, D., and L. A. Smith. "Visualizing bifurcations in high dimensional systems: The spectral bifurcation diagram." *International Journal of Bifurcation and Chaos in Applied Sciences and Engineering* 13, no. 10 (2003).

Ørsted, H.C. *Selected scientific works of Hans Christian Orsted.* Translated by Andrew D Jackson, Karen Jelved, and Ole Knudsen. Princeton, NJ: Princeton University Press, 1998.

Osnaghi, S., F. Freitas, and O. Freire Jr. "The origin of the Everettian heresy." *Studies In History and Philosophy of Science Part B: Studies In History and Philosophy of Modern Physics* 40 (2009): 97–123.

Overbye, D. "Physicists find elusive particle seen as key to universe." *New York Times*, 4 July 2012.

Pacioli, L. *Divina proportione opera a tutti glingegni perspicaci e curiosi necessaria.* Venice: A. Paganius Paganinus, 1509.

Pais, A., and R. Penrose. *Subtle is the Lord: The science and the life of Albert Einstein.* Oxford: Oxford University Press, 2005.

Palladio, A. *The four books of architecture.* Translated by I. Ware. New York: Dover, 1965.

Palmer, J. "Parmenides." *The Stanford encyclopedia of philosophy (Fall 2008 edition).* Edited by E.N. Zalta. 2008. http://plato.stanford.edu/archives/fall2008/entries/parmenides/ (accessed 17 December 2011).

Palmer, T.N. "The invariant set postulate: A new geometric framework for the foundations of quantum theory and the role played by gravity." *Proceedings of the Royal Society A: Mathematical, Physical and Engineering Sciences* 465, no. 2110 (2009): 3165.

Panek, R. *The 4 percent universe: Dark matter, dark energy, and the race to discover the rest of reality.* Boston: Houghton Mifflin Harcourt, 2011.

Pareto, V. *Manuale di economia politica.* Milano: Societa Editrice, 1906.

Pauli, W. "Über den Zusammenhang des Abschlusses der Elektronengruppen im Atom mit der Komplexstruktur der Spektren (On the connexion between the completion of electron groups in an atom with the complex structure of spectra)." *Z. Phys.* 31 (1925): 765.

———. *Writings on physics and philosophy.* Edited by C.P. Enz and K. von Meyenn. Translated by R. Schlapp. Berlin: Springer, 1994.

Peat, F.D. *Superstrings and the search for the theory of everything.* London: Abacus, 1991.

———."Creativity: The meeting of Apollo and Dionysus." *www.fdavidpeat.com.* 25 April 2008. http://www.fdavidpeat.com/bibliography/essays/brusstk.htm (accessed 9 February 2012).

Penrose, R. *The road to reality: A complete guide to the laws of the universe.* London: Vintage, 2005.

Peterson, I. *Newton's clock.* New York: W.H. Freeman, 1993.

Philoponus, J. *Commentary on Aristotle's "Physics."* Edited by H. Vitelli. Berlin: Reimer, 1887.

Planck, M. "Über das Gesetz der Energieverteilung im Normalspektrum (On the law of distribution of energy in the normal spectrum)." *Annalen Der Physik* 4 (1901): 553.

Plato. *The republic bk. VI.* Translated by P. Shorey. Cambridge, MA: Harvard University Press, 1930–1935.

———. *Philebus.* Translated by B. Jowett. New York: Random House, 1937.

———. *Symposium and the death of Socrates.* Ware: Wordsworth, 1997.

———. *The republic.* Translated by B. Jowett. Mineola, NY: Dover Publications, 2000.

Plerou, V., et al. "Scaling of the distribution of price fluctuations of individual companies." *Physical Review E* 60 (1999): 6519–6529.

Ploeger, J. *The boundaries of the new frontier: Rhetoric and communication at Fermi National Accelerator Laboratory.* Columbia, SC: University of South Carolina Press, 2009.

Poincaré, H. *Science and method.* Translated by F. Maitland. New York: Dover, 1952.

Pope, A. "Epitaph intended for Sir Isaac Newton." In *Alexander Pope: A critical edition of the major works*, edited by P. Rogers, 242. Oxford: Oxford University Press, 2006.

Popper, K.R. *The open society and its enemies.* Princeton, NJ: Princeton University Press, 1966.

Pratchett, T. *Soul music.* New York: HarperPrism, 1995.

Quetelet, A. *A treatise on man and the development of his faculties.* Edinburgh: W. and R. Chambers, 1842.

Quinn, S. *Marie Curie: A life.* New York: Simon & Schuster, 1995.

Rand, A. *Anthem.* London: Cassell, 1938.

Randall, L. "Seems ratio of x to y chromosomes hasn't changed in 100 years since first Solvay conference in 1911…" 19 October 2011. http://twitter.com/#!/lirarandall/status/126766144152539137 (accessed 25 October 2011).

Regard, M., and T. Landis. "Beauty may differ in each half of the eye of the beholder." In *Beauty and the brain: Biological aspects of aesthetics*, edited by I. Rentschler, B. Herzberger and D. Epstein. Basel: Birkhauser Verlag, 1988.

Reich, E.S. "Boost for Higgs from Tevatron data." *Nature*, 7 March 2012.

Rhodes, R. *The making of the atomic bomb.* New York: Simon & Schuster, 1986.

Richardson, L.F. *Weather prediction by numerical process.* Cambridge: Cambridge University Press, 1922.

Robinson, J.M. *An introduction to early Greek philosophy.* Boston: Houghton Mifflin, 1968.

Roebber, P. J., D.M. Schultz, B.A. Colle, and D.J. Stensrud. "Towards improved prediction: High-resolution and ensemble modeling systems in operations." *Wea. Forecasting* 19 (2004): 936–949.

Rosen, R. *Essays on life itself.* New York: Columbia University Press, 2000.

Roszak, T. *The gendered atom: Reflections on the sexual psychology of science.* Berkeley, CA: Conari Press, 1999.

Rovelli, C., and L. Smolin. "Loop space representation for quantum general relativity." *Nuclear Physics B* 331 (1990): 80.

Rowell, G. "The power of one—how individuals have succeeded in preserving wilderness." *Sierra*, September 1995: 73.

Rucker, R. "2012: What is your favorite deep, elegant, or beautiful explanation?" *Edge.org.* 16 January 2012. http://www.edge.org/responses/what-is-your-favorite-deep-elegant-or-beautiful-explanation (accessed 17 January 2012).

Rukeyser, M. *The speed of darkness.* New York: Random House, 1968.

Russell, A.S. "F. Soddy, interpreter of atomic structure." *Science* 124 (1956): 1069–1070.

Russell, B. *Mysticism and logic.* London: Longmans, 1898.

Rutherford, E. "The scattering of alpha and beta particles by matter and the structure of the atom." *Phil. Mag.* 21 (1911): 669.

———. *Anniversary address delivered before the Royal Society of London by the President, Sir Ernest Rutherford, November 30th 1927.* London: Harrison, 1927.

Saccheri, G. *Girolamo Saccheri's Euclides vindicatus.* Translated by G.B. Halsted. Chicago: Open Court, 1920.

Sagan, C. *Pale blue dot: A vision of the human future in space.* New York: Ballantine Books, 1997.

Sagan, L. "On the origin of mitosing cells." *Journal of Theoretical Biology* 14, no. 3 (1967): 255–274.

Salam, A. "Elementary particle physics: Relativistic groups and analyticity." In *Proceedings of the eighth Nobel symposium*, edited by N. Svartholm, 367. Stockholm: Almquvist and Wiksell, 1968.

———. *Unification of Fundamental Forces.* Cambridge: Cambridge University Press, 1990.

Sample, I. "The god of small things." *Guardian*, 17 November 2007.

———. *Massive: The hunt for the God particle.* London: Virgin Books, 2010.

———. "Stephen Hawking: 'There is no heaven; it's a fairy story.'" *Guardian*, 16 May 2011a.

———. "Genetics tests flawed and inaccurate, say Dutch scientists ." *Guardian*, 30 May 2011b.

Sampson, S. "2012: What is your favorite deep, elegant, or beautiful explanation?" *Edge.org.* 16 January 2012. http://www.edge.org/responses/what-is-your-favorite-deep-elegant-or-beautiful-explanation (accessed 17 January 2012).

Sato, K. "First-order phase transition of a vacuum and the expansion of the Universe." *Monthly Notices of Royal Astronomical Society* 195 (1981): 467.

Satter, R. "On 70th birthday, Stephen Hawking repeats call to colonize other worlds." *Associated Press*, 8 January 2012.

Scherk, J., and J.H. Schwarz. "Dual models for non-hadrons." *Nuclear Physics B* 81, no. 1 (1974): 118–144.

Schmittner, A., et al. "Climate sensitivity estimated from temperature reconstructions of the last glacial maximum." *Science Express*, November 2011.

Schneider, H. "At IMF, the hunt for a new consensus." *Washington Post*, 7 March 2011.

Schneider, S.H. "Climate science projections." *climatechange.net.* 2005. http://stephenschneider.stanford.edu/Climate/Climate_Science/ClimateScienceProjections.html (accessed 18 December 2011).

Schrödinger, E. "Quantizierung als Eigenwertproblem (Erste Mitteilung) (Quantization as a problem of proper values [part I])." *Annalen der Physik* 79 (1926): 361.

———. *What is life?* Cambridge: Cambridge University Press, 1944.

Schweber, S.S. "2011 Pais Prize lecture: Shelter Island revisited." *American Physical Society.* 2011. http://www.aps.org/units/fhp/newsletters/spring2011/schweber.cfm#1 (accessed 1 January 2012).

Scruton, R. *The aesthetics of music.* Oxford: Oxford University Press, 1999.

Senn, P.R. *The influence of Nietzsche on the history of economic thought.* Vol. 26. In *Research in the history of economic thought and methodology: A research annual*, edited by W.J. Samuels, J. Biddle, and R.B. Emmett. Bingley: Emerald, 2008.

Shackley, S., P. Young, S. Parkinson, and B.E. Wynne. "Uncertainty, complexity and concepts of good science in climate change modelling: Are GCMs the best tools?" *Climatic Change* 38 (1998): 159–205.

Shakespeare, W. *Pericles, Prince of Tyre.* Edited by A.R. Bellinger. New Haven, CT: Yale University Press, 1925.

Shiva, V. *Hidden variables and locality in quantum theory.* PhD thesis. London, ON: University of Western Ontario, 1978.

Sime, R.L. *Lise Meitner: A life in physics.* Berkeley, CA: University of California Press, 1996.

Smolin, L. "Did the universe evolve?" *Classical and Quantum Gravity* 9 (1992): 173–191.

———. *The life of the cosmos.* New York: Oxford University Press, 1997.

———. "Loop quantum gravity." *Edge.org.* 24 February 2003. http://www.edge.org/3rd_culture/smolin03/smolin03_index.html (accessed 21 November 2011).

———. "The case for background independence." In *The structural foundations of quantum gravity*, edited by D. Rickles, S. French, and J.T. Saatsi. Oxford: Oxford University Press, 2006a.

———. *The trouble with physics: The rise of string theory, the fall of a science, and what comes next.* New York: Houghton Mifflin, 2006b.

Snell, C., D. Orrell, E. Fernandez, C. Chassagnole, and D. Fell. "Systems biology approaches to cancer drug development." In *Cancer Systems Biology, Bioinformatics and Medicine*, edited by A. Cesario and F. Marcus, 367–380. London: Springer, 2011.

Snow, C.P., and S. Collini. *The two cultures.* Cambridge: Cambridge University Press, 1993.

Snyder, A. "Autistic genius?" *Nature* 428, no. 6982 (2004): 470–471.

Soddy, F. *Cartesian economics.* London: Hendersons, 1922.

———. *Wealth, virtual wealth and debt. The solution of the economic paradox.* New York: Dutton, 1926.

———. *The role of money: What it should be, contrasted with what it has become.* London: Routledge, 2003.

Specter, M. "The power of nothing." *New Yorker*, 12 December 2011.

Spencer, H. *Social statics, or the conditions essential to human happiness specified and the first of them developed.* London: J. Chapman, 1851.

Sperry, R.W. "Left-brain, right-brain." *Saturday Review*, 9 August 1975: 30–33.

Sperry, R.W., and M.S. Gazzaniga. "Language following surgical disconnection of the hemispheres." In *Brain Mechanisms Underlying Speech and Language*, edited by F.L. Darley, 108–121. New York: Grune and Stratton, 1967.

Stainforth, D. A., M. R. Allen, E. R. Tredger, and L. A. Smith. "Confidence, uncertainty and decision-support relevance in climate predictions." *Philosophical Transactions of the Royal Society of London A* 365 (2007): 2145–2161.

Stainforth, D.A., et al. "Uncertainty in predictions of the climate response to rising levels of greenhouse gases." *Nature* 433 (2005): 403–406.

Stewart, I. *Why beauty is truth: The history of symmetry.* New York: Basic Books, 2007.

Stolnitz, J. "On the origins of 'aesthetic disinterestedness.'" In *Aesthetics: A critical anthology*, edited by G. Dickie and R.J. Sclafani, 607–625. New York: St. Martin's Press, 1977.

Sullivan, A., W.S. Gilbert, and B. Treharne. *The Mikado, or, the town of Titipu.* New York: G. Schirmer, 1940.

Susskind, L. *The cosmic landscape: String theory and the illusion of intelligent design.* New York: Little, Brown and Co., 2006.

Sutton, C. "Steven Weinberg: Master builder of the Standard Model." *Cern Courier*, 25 August 2009.

Talbott, S.L. "From mechanism to a science of qualities." *The Nature Institute.* 9 November 2004. http://www.natureinstitute.org/txt/st/mqual/index.htm (accessed 11 August 2011).

Taleb, N.N. "Errors, robustness, and the fourth quadrant." *International Journal of Forecasting*, 2009: 744–759.

Tarr-Whelan, L. *Women lead the way: Your guide to stepping up to leadership and changing the world.* San Francisco: Berrett-Koehler, 2009.

Taylor, G.I. "Interference fringes with feeble light." *Proc. Cam. Phil. Soc.* 15 (1909): 114.

Taylor, J.B. *My stroke of insight: A brain scientist's personal journey.* New York: Viking, 2008.

Taylor, M.C. *The moment of complexity: Emerging network culture.* Chicago: University of Chicago Press, 2001.

Tegmark, M. "The interpretation of quantum mechanics: Many worlds or many words?" *Fortsch. Phys.*, 1998: 855–862.

———. "Max Tegmark forecasts the future." *New Scientist*, 18 November 2006.

Tett, G. "Could 'Tobin tax' reshape financial sector DNA?" *Financial Times*, 27 August 2009.

The National Academies. "The Shelter Island Conference." *NationalAcademies.org.* 2012. http://www7.nationalacademies.org/archives/shelterisland.html (accessed 1 January 2012).

The National Atomic Museum. *Trinity site: The 50th anniversary of the atomic bomb.* Albuquerque, NM : U.S. Department of Energy, 2008.

The Philosophy Gift Shop. *The Philosophy Gift Shop.* http://www.cafepress.co.uk/philosophy_shop/973063 (accessed 10 September 2011).

Thomasson, L. "Strategists see 17% S&P 500 rise on fed cuts after saying 'Buy.'" *Bloomberg.com.* 5 January 2009. http://www.bloomberg.com/apps/news?pid=newsarchive&sid=a5mSOfz7Alhk&refer=home (accessed 6 January 2009).

Thomson, J.J. "Cathode rays." *Philosophical Magazine* 44 (1897): 293.

———. "On the masses of the ions in gases at low pressures." *Philosophical Magazine* 48, no. 295 (1899): 547–567.

———. *The corpuscular theory of matter.* London: A. Constable & Co., 1907.

Thomson, W. "Nineteenth century clouds over the dynamical theory of heat and light." *Philosophical Magazine*, 1901: 1–40.

Tipler, F. "2012: What is your favorite deep, elegant, or beautiful explanation?" *Edge.org.* 16 January 2012. http://www.edge.org/responses/what-is-your-favorite-deep-elegant-or-beautiful-explanation (accessed 16 January 2012).

Tokyo International Forum. "Future challenges for systems biology: The Tokyo declaration." Tokyo, 2008.

Tolstoy, L. "Art as the communication of feeling: from 'What is art?'." In *Aesthetics: A critical anthology*, edited by G. Dickie and R.J. Sclafani, 53. New York: St. Martin's Press, 1977.

Toth, Z., E. Kalnay, S. Tracton, S. Wobus, and J. Irwin. "A synoptic evaluation of the NCEP ensemble." In *Proc. 5th workshop on meteorological operating systems*, edited by T. Palmer. Shinfield Park: European Centre for Medium-Range Weather Forecasting, 1996.

Triana, P. *Lecturing birds on flying: Can mathematical theories destroy the financial markets?* New York: Wiley, 2009.

Trimmer, J.D. "The present situation in quantum mechanics: A translation of Schrödinger's 'cat paradox' paper." *Proceedings of the American Philosophical Society* 124, no. 5 (1980): 323–338.

Tuan, Y. *Segmented world and self.* Minneapolis: University of Minnesota Press, 1982.

Tversky, A., and E. Shafir. "The disjunction effect in choice under uncertainty." *Psychological Science* 3, no. 5 (1992): 305–9.

Vaucanson, J. *Le mécanisme du fluteur automate.* Edited by C. Cardinal. Translated by M. Hyman. Paris: Ed. des Archives contemporaines, 1985.

Veblen, T. "Why is economics not an evolutionary science?" *The Quarterly Journal of Economics* 12, no. 4 (1898): 373–397.

Veneziano, G. "Construction of a crossing-symmetric, Regge-behaved amplitude for linearly rising trajectories." *Nuovo Cimento A* 57 (1968): 190–197.

Vitali, S., J.B. Glattfelder, and S. Battiston. "The network of global corporate control." *PLoS ONE* 6, no. 10 (2011): e25995.

Vitruvius. *Vitruvius: The ten books on architecture.* Translated by M.H. Morgan. New York: Dover, 1960.

———. *De Architectura.* Translated by Cesare Cesariano (Como, Italy: Gottardo da Ponte for Agostino Gallo and Aloisio Pirovano, 1521).

Voltaire. *Candide: Or, optimism.* Translated by J.E. Butt. London: Penguin, 1947.

Von Franz, M.-L. "The process of individuation." In *Man and his symbols*, edited by C.G. Jung. New York: Dell, 1968.

von Helmholtz, H. "On the conservation of force." In *Scientific memoirs*, edited by J. Tyndall and W. Francis. London: Taylor and Francis, 1853.

Waldrop, M.M. *Complexity: The emerging science at the edge of order and chaos.* New York: Simon & Schuster, 1992.

Walsh, F. "Era of personalised medicine awaits." *BBC.com.* 8 April 2009. http://news.bbc.co.uk/1/hi/health/7954968.stm (accessed 3 January 2012).

Wang, J., S. Zhang, Y. Wang, L. Chen, and X-S. Zhang. "Disease-aging network reveals significant roles of aging genes in connecting genetic diseases." *PLoS Comput. Biol.* 5, no. 9 (2009): e1000521.

Watson, J.D. *The double helix: A personal account of the discovery of the structure of DNA.* London: Weidenfeld and Nicolson, 1968.

Watson, J.D., and F.H.C. Crick. "Molecular structure of nucleic acids: A structure for deoxyribose nucleic acid." *Nature* 171 (1953): 737.

Webb, R. "Dirac equation: a relative success." *Nature*, 28 February 2008.

Weinberg, S. "A model of leptons." *Phys. Rev. Lett.* 19 (1967): 1264–1266.

———. "Conceptual foundations of the unified theory of weak and electromagnetic interactions." *Rev. Mod. Phys.* 52 (1980): 515–523.

———. "Newtonianism, reductionism and the art of congressional testimony." *Nature* 330 (1987): 433–437.———. *Dreams of a final theory.* New York: Pantheon Books, 1992.

———. *Facing up: Science and its cultural adversaries.* Cambridge, MA: Harvard University Press, 2001.

Weiner, C. *Interview with Sir James Chadwick.* Cambridge, 20 April 1969.

Weintraub, E.R. *How economics became a mathematical science.* Durham, NC: Duke University Press, 2002.

Wertheim, M. *Pythagoras' trousers: God, physics, and the gender wars.* London: Fourth Estate, 1997.

———. "Numbers are male, said Pythagoras, and the idea persists." *New York Times*, 3 October 2006.

———. *Physics on the fringe: Smoke rings, circlons, and alternative theories of everything.* New York: Walker, 2011.

Wesson, R. *Beyond natural selection.* Cambridge, MA: MIT Press, 1994.

Weyl, H. "Reine Infinitesimalgeometrie." *Math. Z.* 2 (1918a): 384–411.

Weyl, H. "Gravitation und Elektrizitat." *Sitzungsber. Preuss. Akad. Wiss.*, 1918b: 465–480.

———. *Symmetry.* Princeton, NJ: Princeton University Press, 1952.

Wheeler, J.A. "How come the quantum?" *Annals of the New York Academy of Sciences* 480 (1986).

Whitehead, A.N. *The concept of nature, Tarrner lectures delivered in Trinity College, November, 1919.* Cambridge: The University Press, 1920.

———. *Science and the modern world.* New York: Macmillan Co., 1925.

Whitrow, G.J. "Einstein: The man and his achievement." Edited by G.J. Whitrow. New York: Dover, 1973.

Wiener, N. *God and Golem, Inc.* Cambridge, MA: MIT Press, 1964.

Wigner, E. "The unreasonable effectiveness of mathematics in the natural sciences." *Communications in Pure and Applied Mathematics* 13, no. 1 (1960).

Wilson, R.R. *R.R. Wilson's congressional testimony, April 1969.* 17 April 1969. http://history.fnal.gov/testimony.html (accessed 9 December 2011).

———. *Starting Fermilab: Some personal viewpoints of a laboratory director (1967–1978).* Fermilab Annual Report. Chicago: Fermilab, 1987.

Witten, E. "String theory dynamics in various dimensions." *Nucl. Phys. B* 443 (1995): 85–126.

———. "Reflections on the fate of spacetime." *Physics Today* 49, no. 4 (April 1996): 24–30.

Wittgenstein, L. *Culture and value.* Edited by G.H. von Wright and H. Nyman. Chicago: University of Chicago Press, 1980.

Wohlforth, C. *The whale and the supercomputer: On the northern edge of climate change.* New York: North Point Press, 2004.

Woit, P. *Not even wrong: The failure of string theory and the search for unity in physical law.* London: Vintage, 2006.

Wolchover, N. "Can a butterfly in Brazil really cause a tornado in Texas?" *Life's Little Mysteries,* 12 December 2011. http://www.lifeslittlemysteries.com/1989-butterfly-effect-weather-prediction.html (accessed 12 December 2011).

Wolfram, S. *A new kind of science.* Champaign, IL: Wolfram Media, 2002.

Wood, G. *Living dolls: A magical history of the quest for mechanical life.* London: Faber and Faber, 2002.

Woods, A. *Medium-range weather prediction: The European approach; The story of the European Centre for Medium-Range Weather Forecasts.* New York: Springer, 2005.

Worthington, A.M. *The splash of a drop: Being the reprint of a discourse delivered at the Royal Institution of Great Britain, May 18, 1894.* London: Society for Promoting Christian Knowledge, 1895.

Wroe, N. "David Hockney: A life in art." *Guardian*, 13 January 2012.

Young, T. *On the nature of light and colours.* Vol. 1 in *A Course of Lectures on Natural Philosophy and the Mechanical Arts*, 359. London: Joseph Johnson, 1807.

Zaidel, D.W., and C. Deblieck. "Attractiveness of natural faces compared to computer constructed perfectly symmetrical faces." *Intern. J. Neuroscience* 117 (2007): 423–31.

Zee, A. *Fearful symmetry: The search for beauty in modern physics.* Princeton, NJ: Princeton University Press, 1986.

Zetter, K. "Surfer-physicist's unified theory leads to fame, backlash." *Wired*, 27 February 2008.

Zöllner, F., and J. Nathan. *Leonardo da Vinci: The complete paintings and drawings.* Köln: Taschen, 2003.

Index